OPTIMIZATION
METHODS

with Applications
for Personal
Computers

TERRY E. SHOUP

Florida Atlantic University

FARROKH MISTREE

University of Houston

PRENTICE-HALL, INC., Englewood Cliffs, New Jersey 07632

Library of Congress Cataloging-in-Publication Data

Shoup, Terry E., (date)
  Optimization methods.

  Includes bibliographies and index.
  1. Mathematical optimization—Data processing.
  2. Microcomputers—Programming. I. Mistree, Farrokh.
  II. Title.
  QA402.5.S5434 1987   620′.00425′0285526   86-17069
  ISBN 0-13-638172-3

Editorial/production supervision: *Lynn Frankel*
Manufacturing buyer: *Gordon Osbourne*

Printed in the United States of America
10 9 8 7 6 5 4 3 2 1

ISBN 0-13-638172-3

Prentice-Hall International (UK) Limited, *London*
Prentice-Hall of Australia Pty. Limited, *Sydney*
Prentice-Hall Canada Inc., *Toronto*
Prentice-Hall Hispanoamericana, S.A., *Mexico*
Prentice-Hall of India Private Limited, *New Delhi*
Prentice-Hall of Japan, Inc., *Tokyo*
Prentice-Hall of Southeast Asia Pte. Ltd., *Singapore*
Editora Prentice-Hall do Brasil, Ltda., *Rio de Janeiro*

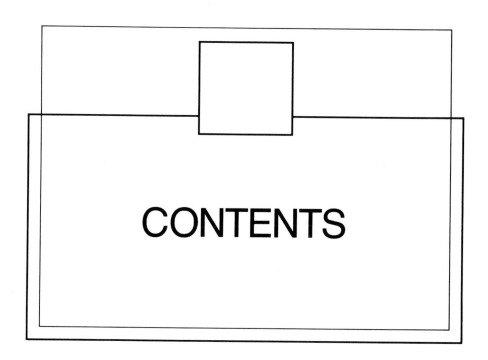

CONTENTS

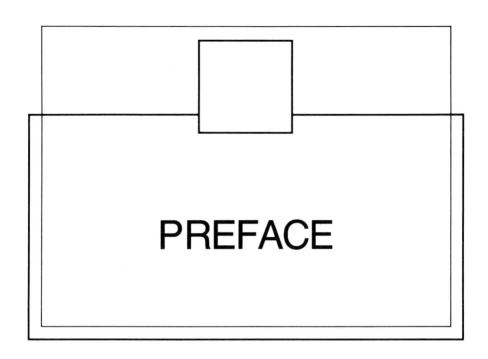

PREFACE

The personal computer is a truly remarkable tool for digital manipulation. Recent growth trends in the area of personal computer technology predict the increased utilization of these valuable tools in many aspects of our lives in the decades ahead. Yet with all of its widespread acceptance in the public domain, the personal computer has seen relatively little use in applications for numerical methods in engineering design. A primary reason for this fact is that most numerical methods texts and software packages are written from the perspective of the large, mainframe computer. Yet the use of the small computer as a tool for implementing certain types of engineering design problems is rapidly becoming preferable to the larger computer for reasons of cost, utility, and convenience. Of the limited information and software that is available for implementing design methods on personal computers, very little has been developed by those who have a combined appreciation for the power of modern optimization methods and the versatility of the personal computer. It is the purpose of this book to exploit the best characteristics of these combined domains. It is intended to provide not only a source of information about the area of design optimization, but also to provide a software data base containing high-quality algorithms tailored for implementation on the personal computer. The book and its accompanying software should provide a package suitable for teaching design optimization courses at universities and suitable for individual engineering personnel who want to expand the utility of their computational equipment.

The basic approach of this book is to move from the fundamentals of optimization through to those advanced topics encountered in engineering design optimization. As the methods are developed, corresponding optimization software is also presented in the form of subroutine software together with example applications.

A summary is presented at the end of each chapter to help the user choose the best algorithm for a given optimization task and to alert the user to potential problems in the application of these algorithms on the personal computer. End-of-chapter exercises are provided for use by the reader, and end-of-chapter references are provided to help the user discover more information about specific algorithms. For those who find the BASIC subroutine package used in this text to be useful, a software package based on this software together with a user's guide is available. (See page 168 for ordering information.) Also, for those who wish to probe more deeply into the advanced topics of optimization, an advanced text is in preparation.

The authors would like to thank all of those who contributed to the construction of this text and the approach it provides. Special thanks for encouragement are due to Dr. R. H. Page. A special note of appreciation is due to those of our colleagues who provided helpful suggestions to improve the text.

Terry E. Shoup
Farrokh Mistree

1

INTRODUCTION

Advances in VLSI technology have helped the microcomputer to achieve a significant role in the problem-solving environment of science and engineering. (Courtesy of Intel.)

We live in a world of products described in terms of comparatives and superlatives. To be regarded as successful, a product must be cleaner, faster, more powerful, more attractive, more efficient, or more cost-effective than its competitors. Success in our society is most frequently measured in terms of a comparison with competing products and processes. Rewards and acclaim are offered to the inventors and developers who succeed in producing the best products and processes. As our society transforms itself from an industrial society into an information society, the principle of improved products and processes is even more pronounced. Constraints on the design process become more severe, and the need for a competitive edge can easily spell the difference between success and failure. For this reason, it is especially important for the engineering designer to have available the necessary informational tools to compete successfully in this rapidly accelerating environment. Optimization is one such tool.

In the past, optimization was discussed only within the context of large mainframe computers. Recently, however, with the rapid growth and development of the capabilities of personal computers, the numerical tools of optimization are now becoming easily available to every engineering designer. The joining together of the microcomputer and the field of optimization is a merger whose time is right and whose potential is great. It is the purpose of this book to serve as a basic information resource regarding optimization methods for those who wish to apply the computational power of the personal computer to the field of engineering design. This text focuses on three important goals:

1. To describe the design process in a way that will explain how optimization methods can be useful
2. To describe the spectrum of optimization methods from fundamentals through to advanced topics, with a view toward identifying those methods that are most useful in engineering design
3. To present practical computational algorithms that provide an efficient merger of state-of-the-art optimization techniques with personal computer capability

1-1 THE PERSONAL COMPUTER IN ENGINEERING PROBLEM SOLVING

As we undertake this study, it is assumed that the reader has a modest knowledge of computer fundamentals and a working understanding of how to use a microcomputer. The reader who is deficient in either of these topic areas may wish to invest some time in studying them before proceeding.

Before initiating a study of the field of optimization from the perspective of the personal computer, it is well to begin by discussing two important issues that are fundamental to this study. These are:

1. Why should one consider using the personal computer to handle engineering problems?

2. What are the limitations of most concern when considering the use of the personal computer to solve an engineering problem?

The answer to the first question is found in considerations of cost and utility. Quite simply, there are engineering problems that should be solved on the personal computer because it is more cost-effective and because it is more convenient. Let us look at the evidence of these two reasons.

Cost

It can be shown that a personal computer is about an order of magnitude less expensive per byte of storage to operate than a mainframe computer. It can also be shown that the CPU processing time for a large computer is only an order of magnitude faster than that for a microcomputer, while the overall cost is more than three orders of magnitude. Thus the overall cost per CPU cycle time is overwhelmingly in favor of the smaller computer. In addition, since the microcomputer does not require specially trained operators or specially prepared operating environments, its overall operational cost may be far more favorable than that predicted from a comparison of purchase price. Because it is based on relatively new technologies, it is likely that the purchase price of microcomputer systems will continue to decrease in the future. This fact adds to the already strong economic reasons for using the microcomputer for computational tasks within its utility range.

Utility

One of the major differences between a large computer and a small computer is in the number of users who can simultaneously interact with the machine. Under the multiprogramming environment of a large computer several hundred users can use this computational resource simultaneously. On the other hand, for the microcomputer, operating systems are designed so that the machine functions entirely for a single user. For this reason, microcomputers are referred to as "personal" computers. It is this capability for dedicated service that makes the personal computer such an appealing tool for the engineering user. Indeed, with the recent increased availability of *portable* personal computers, the applications potential of these devices has generated an even greater appeal for their use.

In the area of computer languages supported, most large-scale computers support a wide variety of commonly used high-level languages. Personal computers, on the other hand, frequently support only one higher-level language. In most cases this is the BASIC language. Although microcomputers can usually be programmed in machine language, this process is different from machine to machine and makes

Figure 1-1 Because of its relatively low cost and high performance, the personal computer has become a significant problem-solving tool for business, education, science, and engineering. (Courtesy of IBM.)

transporting software somewhat difficult. Most scientific and engineering problem solving is presently accomplished using the FORTRAN language. Because of the widespread popularity of the microcomputer, special versions of this language are becoming available. Nevertheless, there remains a strong need for engineering software in the BASIC language. Over the past decade a number of excellent optimization packages have been developed in FORTRAN for use on a variety of large computers. There is every reason to believe that the availability of similar packages in BASIC for implementation on the personal computer will begin to emerge in the near future. Thus it would seem that the best advantages of personal computer utility and software versatility should provide a synergistic thrust to the field of optimization. It is hoped that this book will provide a significant first step in helping this to become a reality.

Let us turn now to a discussion of the second important issue when considering the topic of optimization for the personal computer.

Limitations to Consider when Using the Personal Computer

The second question raised concerning the use of the personal computer in engineering problem solving relates to the limitations of this device. There are four limiting parameters to consider in deciding whether to use the personal computer in any problem-solving setting. These are as follows:

Time of Run. Because the small computer runs an order of magnitude slower than its larger counterpart, the small computer should be limited to applications that can be completed in a reasonable time.

Space. Because of the limited storage space and word size available through small computers, these devices should be used for solving problems that do not require large amounts of computational storage space.

Accuracy. Because most small computers operate with 8-bit or 16-bit words, most microcomputers require several words to represent a single floating-point value. Thus the degree of accuracy available from a given personal computer may differ from machine to machine. For applications requiring higher accuracy than that available from a given microcomputer, the larger computer should be used.

Amounts of I/O. Because of their relatively slow input and output capabilities, small computers should be utilized for problems requiring modest amounts of input and output information.

1-2 THE DESIGN PROCESS

Numerical optimization is a way to automate the engineering design process. Before we look at the techniques for optimization, we must look at the nature of the design process itself. Although a number of authorities on the methodology of design have presented descriptions of the process, most of the descriptions tend to be similar. The design process, as we will describe it, involves the six-step procedure diagrammed in Figure 1-2. The steps are represented by boxes in the diagram, and arrows connecting the boxes indicate the seqential order in which they are applied. We will look at the activity that takes place during each of these steps.

Step 1: Recognize the Need

Some people mistakenly believe that engineers create need. This is, of course, no more true than the notion that doctors create illness or that farmers create hunger. The products and processes created by engineering design are a direct response to specific needs of society. The logic for this cause-and-effect relationship is obvious. Engineers cannot make a living in the creation of products and processes for which there is no societal need. The ability to recognize present needs and to anticipate future needs is an extremely valuable talent for an engineer to cultivate.

The first step in the design process, the recognition of need, is probably the single most important part of the overall process. Yet it is frequently given inadequate treatment in the thought process of design. It is also a step that is not well suited for automated assistance by the personal computer. The engineering designer needs to make a deliberate effort to focus attention on the preparation of a suitable statement of need for the problem that he or she is undertaking. A carefully formulated statement of need can often save considerable time and energy later in the

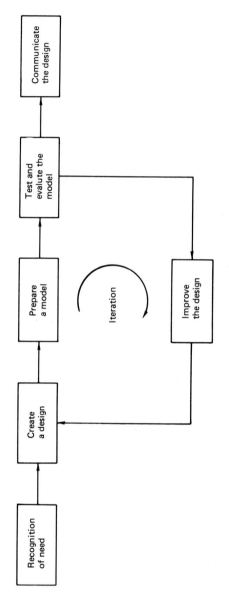

Figure 1-2 The design process.

design cycle. Implied in any statement of need is an identification of the real constraints on the problem being considered. A statement of need that recognizes and describes the fundamental elements of a problem can often keep the designer from imposing artificial constraints on his or her design activities. For example, a group of engineers was once engaged in the design of a new type of eyeglasses with expanded visibility and reduced weight. When the design group finally recognized and restated the true need to be that of achieving an improved method of vision correction, they were freed from the artificial constraint of requiring their output to be eyeglasses. The result was the development of the first pair of contact lenses. Until they were able to see the fundamental elements of their design task, they unnecessarily limited their perspective to solutions of low innovation. Some of the most exciting and innovative progress occurs in quantum technological jumps rather than in continuous improvements. For example, automobile manufacturers have worked for years to improve motor oils so that they will need to be replaced less often. A careful assessment of the fundamental elements of this need may reveal that what is actually needed is an engine that does not require a liquid lubricant. A quantum jump in technology may be possible if design efforts are diverted from motor oil design to the design of the engine itself.

It is said that the only way to know that progress has taken place is to be able to measure it. Thus a statement of need should contain adjectives that imply quantitative and qualitative measures of how well a given design satisfies the need. For example, in the case of the illustration of the contact lens design, the measure of quality is based on the amount of weight reduced and on the degree of enhanced vision. This ability to measure the merit of a particular design is an essential quality in the field of optimization.

Once a statement of need is established, the designer would be well advised to review the statement periodically during the design process. This review should be used to see if progress is being made in meeting the need and also should provide an opportunity for the designer to see if a clearer statement of need is now apparent.

Step 2: Create a Design

Once the need has been clearly recognized and stated in a succinct way, the next step is to begin creating design ideas that will satisfy this need. Of all the steps in the design process, this step requires the most ingenuity and imagination. Although every person is born with a certain amount of creative talent, much of our natural creativity is suppressed by our social environment. For example, many people are reluctant to suggest a new idea because they fear that someone will laugh. As a result, these people tend to suppress their basic creativity and thus become less and less productive in their work. Fortunately, there are ways of overcoming these suppressive influences so that productive, innovative thought can take place. The overall objective in the design process is to identify a good idea or combination of ideas to meet the stated need. Not surprisingly, the idea or creation step may require a number of attempts in order to achieve an optimum result. Sometimes it works well to modify existing ones. For example, in the automotive industry new design ideas take the form of modifications of existing equipment. In the toy industry, on the other hand, new design ideas take the form of completely new and different products.

Step 3: Prepare a Model

Once an idea has been created, it becomes necessary to find a means to evaluate the quality of that design in satisfying the need requirements. One way to do this, of course, would be to build the suggested design idea. This procedure is usually impractical for reasons of cost, time, and effort. To conserve cost, time, and effort, engineers frequently make use of a simplified model to evaluate a design idea. A model may be real or abstract and may be anything from a simple mental image of the idea to an elaborate mathematical or physical reproduction of the proposed concept. For example, an engineer designing a new type of tennis racket might make a model in the form of a simple sketch of the proposed idea. The sketch will show the important features of the new design and will allow the engineer to make changes or improvements by using only a pencil and an eraser. This kind of model is frequently used in engineering design. There are many other types of models available to the engineer, and each has advantages and limitations. The ability to work with a variety of models is a talent that every designer should cultivate. The personal computer has unique abilities to help the designer with the preparation of a design model (see Figure 1-3).

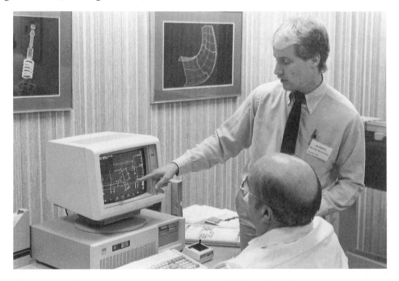

Figure 1-3 The personal computer has unique abilities to help the designer to prepare and visualize a design model. (Courtesy of Manufacturing and Consulting Services, Inc.)

Step 4: Test and Evaluate the Model

Once the model has been prepared, it is time to evaluate the proposed design idea by evaluating the performance of the engineering model. In this manner the engineer has a chance to observe how well the proposed design idea satisfies the required need.

The testing and evaluation of a model may be very simple or quite involved. For example, in the case of the sketch of the new tennis racket, the testing and evaluation may consist entirely of the mental process of deciding whether the design looks strong enough or if it looks functional enough. If the model is a mathematical formula that describes some important characteristics of the proposed idea, the engineer will probably put numbers into the formula to see what quantitative performance the model predicts. For example, in the case of the tennis racket design, if the formula is one that predicts the stresses in the racket handle based on the applied forces, the engineer could use the expected forces in the model to determine if the handle will break due to the stresses.

Frequently, a model will be mathematically complex. When this happens, the designer may choose to use a computer to assist with the numerical testing. In this case the evaluation of the computer output is still the responsibility of the designer. In the design process there is no substitute for the application of engineering judgment.

Step 5: Improve the Design

As a result of the tests performed on the model, the engineer should have a quantitative measure of the success or failure of the idea. The engineer will probably know whether the idea should be abandoned or whether it should be retained for further improvement. One of the fortunate results of the testing and evaluation step is that this process often provides considerable insight into where improvements can and should be made. For example, in the case of the tennis racket design, if the model reveals that the handle will break during normal use, the logical remedy would be to increase the strength of the handle by adding more material at the locations of high stress. It is fortunate that even the poorest design idea will probably give useful information about how to choose a better design. Since a number of different design ideas may be tried, modified, and improved before a final design choice is made, the design cycle of Figure 1-2 tends to be quite iterative. The diagram illustrates this as a curved arrow depicting the iterative flow of this feedback process. In actual practice a single design problem may require from one to a hundred cycles to complete the design process. This suggests that both patience and perseverance are useful attributes for engineers. A well-known illustration of this is to be found in the records of Thomas Edison, who describes trying several hundred ideas before perfecting his design of the light bulb.

Step 6: Communicate the Design

No matter how well a design may satisfy a particular human need, it cannot be converted into a useful product or process if the details of the design are not communicated to those who will implement its use. The communication step thus links the design cycle to the next stage of engineering activity as the idea moves from conceptualization to utilization. Communication of engineering ideas can be by written words, spoken words, or by pictures, graphs, and drawings.

As we think about applying the engineering design methods diagrammed in Figure 1-2, several important facts are worth keeping in mind. First, the human mind is capable of handling straightforward design decisions with remarkable speed. Thus the complete process of making a good engineering design decision need not take excessive time for routine problems. Second, the human mind often has difficulty in defining boundaries between the steps in the design process. Thus the designer may combine one or more of the blocks in the diagram into what seems like a single activity. Although this procedure is not wrong, it should be avoided if it results in inadequate treatment of each important step.

1-3 THE PERSONAL COMPUTER IN ENGINEERING DESIGN

The various steps in the design process often consume different amounts of time and effort. The boundaries between the steps are sometimes difficult to define. Frequently, two or more steps can be combined. The key to knowing when to use the personal computer in the design environment is to recognize its abilities and the ability of the human mind. The computer is best equipped to handle mundane, repetitive tasks and the human mind is best equipped to handle those tasks involving abstract manipulation. Thus those tasks early in the design process are poorly suited to computer assistance and those at the end are best suited. An illustration of this situation is presented in Table 1-1. From the table it is obvious that engineering design optimization has strong potential for microcomputer assistance. There are three basic criteria that must be satisfied if traditional optimization methods are to be

TABLE 1-1 THE POTENTIAL FOR COMPUTER ASSISTANCE IN DESIGN

Aspect	Computer Potential	
Recognition of need	Desirable	but few applications exist
Create a solution	Limited	computer-augmented creativity
Prepare a model	Some	special languages and applications packages
Test and evaluate	Yes	through numerical methods
Improve the design	Yes	optimization
Communicate the design	Yes	computer graphics, computer-aided manufacturing, and word processing

used to assist with any engineering problem. These relate to the fact that the digital computer can handle only numerical information. The criteria are as follows:

1. The design must be described in terms of *numerically measurable parameters*.
2. The design must be capable of evaluation with a *quantifiable measure of merit* in terms of the parameters that define it.
3. The *constraints* on the design must be capable of representation in a *numerical form* in terms of the parameters that define the design.

The structure of these criteria leads us to expect that there must be a body of knowledge that defines in a precise way the vocabulary of optimization. This is indeed true and is the topic of the next chapter.

REFERENCES

1. Gibson, J. E., *Introduction to Engineering Design,* Holt, Rinehart and Winston, New York, 1968.
2. Shoup, T. E., *A Practical Guide to Computer Methods for Engineers,* Prentice-Hall, Inc., Englewood Cliffs, N.J., 1979.
3. Shoup, T. E., *Numerical Methods for the Personal Computer,* Prentice-Hall, Inc., Englewood Cliffs, N.J., 1983.
4. Shoup, T. E., Fletcher, L. S., and Mochel, E. V., *Introduction to Design with Graphics and Design Projects,* Prentice-Hall, Inc., Englewood Cliffs, N.J., 1980.
5. Woodson, T. T., *Introduction to Engineering Design,* McGraw-Hill Book Company, New York, 1966.

2

FUNDAMENTALS OF OPTIMIZATION

Recent advances in computer-aided design and drafting packages have greatly enhanced the utility of the personal computer in the engineering environment. (Courtesy of T & W Systems, Inc.)

Simply stated, optimization provides a logical method for the selection of the best choice from among all possible designs that are available. In recent years this topic has received considerable treatment in the literature, and a number of excellent computer-aided optimization algorithms are available. In this chapter we first define the terminology of optimization and then we discuss the fundamental forms of optimization problems. Next, we discuss the various methods of classification of optimization problems, since this can often provide insight into the choice of a solution algorithm. Finally, we discuss the rationale on which the basic algorithms are based.

2-1 INTRODUCTION TO OPTIMIZATION

The term *optimization,* as it is used in the design literature, is defined to be the process or rationale that a designer uses to achieve an improved solution. Although it is desirable to have the very best or "optimum" solution to a problem, the engineer usually must settle for improvement rather than perfection in designs. For this reason we speak of optimization as the process of *movement toward improvement* rather than achievement of perfection.

If we consider a system defined in terms of m equations and n unknowns, three fundamental types of problems emerge.

1. If $m = n$, the problem is said to be *algebraic* and usually has at least one solution.
2. If $m > n$, the problem is said to be *overconstrained* and is impossible to solve in general.
3. If $m < n$, the problem is *underconstrained* and many usable solutions exist that satisfy the requirements.

In design, the third class of problems occurs most frequently and is generally regarded as the most interesting because it allows the designer to utilize his or her creative talents to select additional constraints or criteria on which to base the design choice. In choosing from among the many usable solutions that exist, the product or process that performs better than its competitors will usually achieve the most success in the marketplace. Thus the importance of optimization in design becomes obvious.

To proceed with our discussion of the topic of optimization, we need a few basic definitions.

Design Variables

The term *design variable* is used to describe the individual elements in a group of independent, variable parameters that uniquely and completely define the design problem being considered. The design variables are unknown values to be solved in the optimization process. They take the form of any of the fundamental or derived units that can be used to quantify engineering systems. For example, they may represent unknown values of length, mass, time, temperature, and so on. The actual number of design variables will determine the complexity and versatility of a given design problem. We generally identify the number of design variables by the integer n and the design variables become x values having subscripts from 1 to n. Thus the n design variables in a given problem will be:

$$x_1, x_2, x_3, \ldots, x_n$$

Merit Function

The *merit function* is an equation or expression that the designer desires to maximize or minimize. It provides a quantitative means for evaluating and comparing the relative quality of two competing designs. Mathematically, the merit function defines an $(n + 1)$-dimensional surface. The value of the merit function will depend on the values of the design variables:

$$M = M(x_1, x_2, \ldots, x_n)$$

Examples of common engineering merit functions to be maximized or minimized are cost, weight, strength, size, and efficiency. If there is only one design variable, the merit function can be plotted as a graph in two dimensions, as shown in Figure 2-1. If there are two design variables, the merit surface can be plotted as a three-dimensional surface, as shown in Figure 2-2. The merit surface for three or more design variables is called a *hypersurface* and cannot be visualized in the traditional

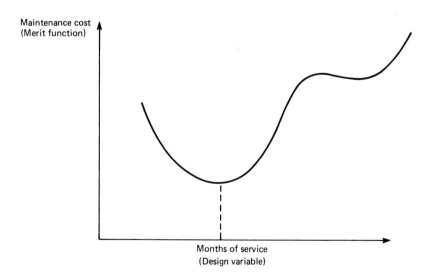

Figure 2-1 A one-dimensional merit function.

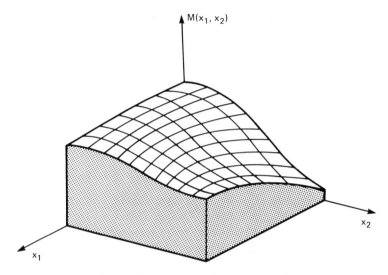

Figure 2-2   A two-dimensional merit function.

sense. The physical and mathematical characteristics of the merit surface are of great importance in the optimization process since the nature of this surface will influence the selection of the most efficient algorithm.

The merit function, as we have defined it, can assume some unusually challenging forms in a given design situation. For example, the merit function sometimes cannot be expressed in closed mathematical form or may involve a piecewise continuous function. Evaluation of the merit function may sometimes require the use of a table of engineering information such as the steam tables or may rely on data gathered in an experiment. For some types of design problems the design variables can assume only integer values rather than real values. Examples of this type of design variable include such things as the number of teeth on a gear or the number of bolts on a flange. Occasionally, the design variables can assume only a value of yes or no rather than a number. Qualitative factors such as customer satisfaction, safety, and aesthetic appeal are difficult to use in optimization schemes since they are cumbersome to quantify numerically. Regardless of the form it takes, however, the merit function must be a unique function of the design variables.

Some types of optimization problems can be formulated in terms of more than one measure of merit. Sometimes these may be in conflict. An example of this situation occurs in aircraft design, where it is desired to maximize strength, minimize weight, and minimize cost. Whenever the possibility for more than one measure of merit exists, the designer must establish priorities and assign weighting values to each measure of merit. This process results in what is called a *trade-off function* and provides a single composite merit value to be used in the optimization process.

Minimization and Maximization

Some optimization algorithms are set up to search for maxima, whereas others are set up to search for minima. Regardless of which type of extremum problem is being solved, a general algorithm may be used since a minimization problem can be converted into a maximization problem simply by changing the sign of the merit value. This situation is illustrated in Figure 2-3.

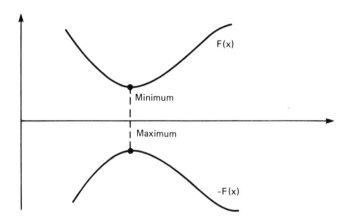

Figure 2-3   Reversing the sign of a merit function will convert a maximization problem into a minimization problem.

Design Space

The total domain defined by all the n design variables is called the *design space*. This space is not as large as one might think since it is usually limited by constraint bounds imposed by the reality of the problem. Indeed, it is possible to constrain a problem so much that no usable design exists. The two categories of constraints are called functional constraints and regional constraints.

Functional Constraints

Functional constraints are functional relationships of the design variables that must be satisfied in the design solution. These arise due to such things as the laws of nature, economics, law, taste, and available materials. In general, there may be any number of these of the form

$$C_1(x_1, x_2, \ldots, x_n) = 0$$
$$C_2(x_1, x_2, \ldots, x_n) = 0$$
$$\cdot$$
$$\cdot$$
$$\cdot$$
$$C_j(x_1, x_2, \ldots, x_n) = 0$$

If any of these functional constraints can be used to solve for one of the design variables as a function of all others, this new expression can be used to eliminate that design variable in the optimization process. In this way the degree of dimensionality of the problem will have been reduced. This procedure is desirable since it will generally reduce the complexity of the problem.

Regional Constraints

Regional constraints are special types of functional constraints that take the form of inequalities. In the most general case, any number of these may exist in the form

$$z_1 \le r_1(x_1, x_2, \ldots, x_n) \le Z_1$$
$$z_2 \le r_2(x_1, x_2, \ldots, x_n) \le Z_2$$
$$\vdots$$
$$z_k \le r_k(x_1, x_2, \ldots, x_n) \le Z_k$$

It should be noted at this point that often an optimum value of a merit function does not occur where the gradient of the merit surface is zero. The best design selection often occurs along one of the constraint boundaries.

Local Optimum

The *local optimum* is a point in the design space that is higher than all other points within its immediate vicinity. Figure 2-4 illustrates a one-dimensional merit function that has two local optima. The designer must not fall prey to selecting the first optimum value found.

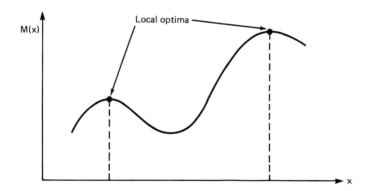

Figure 2-4 A general merit function can have more than one local optimum.

Global Optimum

The *global optimum* is the optimum design within the total allowable design space. It is the best of all local optima and is the design choice that the designer ultimately seeks to find. It is possible for several equal global optima to occur at two or more places in the design space.

The formulation of optimization problems can best be illustrated by means of an example problem.

EXAMPLE 2-1

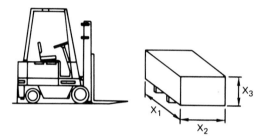

It is desired to design a closed rectangular container to hold 1 m³ of a loose fiber product. It is desired to utilize minimum material (i.e., minimum surface area assuming that the same wall thickness is to be used for every surface of the container) for reasons of cost. In order to allow the container to be carried conveniently by a lift truck, the width of the container must be no smaller than 1.5m. Formulate this problem in a form suitable for treatment by an optimization algorithm.

For this problem, the *design variables* are

$$x_1 \quad x_2, \quad \text{and} \quad x_3$$

the *merit function* to be minimized is the surface area,

$$A = 2(x_1x_2 + x_2x_3 + x_1x_3) \quad \text{m}^2$$

the *functional constraint* is

$$\text{volume} = 1.0 \text{ m}^3 = x_1x_2x_3$$

and the *regional constraint* is

$$1.5 \leq x_1$$

The careful designer will notice that the degree of dimensionality of this problem can be reduced by virtue of the simplicity of the functional constraint. Since

$$x_3 = \frac{1}{x_1 x_2}$$

the need for the design value x_3 can be eliminated to get a new problem formulation.

In the new problem formulation the *design variables* are

$$x_1 \quad \text{and} \quad x_2$$

the *merit function* to be minimized is

$$A = 2(x_1 x_2 + \frac{1}{x_1} + \frac{1}{x_2}) \quad \text{m}^2$$

the *functional constraint* is

$$\text{none}$$

and the *regional constraint* is

$$1.5 \le x_1$$

Once the problem is formulated in this standard form it is ready to be solved by whatever method the engineer chooses to use. The first impulse of the engineer might be to apply a traditional calculus approach and set

$$\frac{\partial A}{\partial x_1} = \frac{\partial A}{\partial x_2} = 0$$

This method will yield $x_1 = x_2 = x_3 = 1.0$. Unfortunately, this solution violates the regional constraint and is therefore not an acceptable design solution. This example serves to illustrate an important fact about optimization: that *due to the constraints on a given problem, the optimum solution may occur at a point other than where the local gradient is zero.*

Complete solution to this example problem is possible by the methods presented in subsequent chapters.

2-2 PENALTY FUNCTION

Nearly all of the optimization techniques that are currently in use can be categorized as methods for unconstrained minimization or maximization. This means that they find an optimum without regard to any constraints. In only a few cases will the actual optimum occur at a point in the design space where the local gradient is zero. Generally, the optimum will occur along one of the constraint boundaries. Since most engineering problems have a number of equality and inequality constraints, the use of an unconstrained optimization algorithm by itself will usually not provide the best approach to finding a usable solution. Yet many of the unconstrained optimization algorithms are quite powerful and offer great potential for use in the design process. Therefore, it seems reasonable to look for some way to adapt constrained optimization problems for solution by unconstrained methods. The mathematical device for accomplishing this task is the *penalty function*.

The basis for the penalty function method is a new merit function of the form

$$M(x_i) = F(x_i) + P(G_j(x_i))$$

In this expression

$F(x_i)$ = merit function (constrained)
$M(x_i)$ = composite merit function (unconstrained)

$P(G_j(x_i))$ = penalty function based on the equality and inequality constraints

The composite merit function is formed by summing the penalty function and the previous objective function. As an example of a penalty function, consider the use of a function $P(G_j(x_i))$ which is zero for all design points satisfying the constraints $G_j(x_i) \geq 0$ and is infinite for all design points that violate the constraints. Clearly, if all constraints are satisfied, the minimization of $M(x_i)$ is equivalent to the minimization of $F(x_i)$. If, on the other hand, any constraint is violated, the merit function goes to infinity, which is far from the minimum of $F(x_i)$. Thus any design that violates the constraints is said to be *penalized*.

There do exist certain difficulties associated with the implementation of the penalty function suggested previously, owing to the harsh nature of the discontinuity of the penalized merit surface at the boundaries. To overcome this situation, a few researchers have shown that a gradual sequence of penalty function minimizations is far more desirable. Thus, instead of solving only one unconstrained optimization problem, a sequence of problems is considered, each of which comes closer to the final desired solution. The Fiacco–McCormick method suggests the use of a penalty function of the form

$$P[G_j(x_i)] = r_p G_j^{-1}(x_i) \qquad p = 1, 2, \ldots$$

where r_p is the penalty parameter and the index p identifies the successive values of the penalty parameter for the sequence of problems considered. Thus the unconstrained merit function becomes

$$M(r_p, x_i) = F(x_i) + r_p \sum_{j=1}^{J} G_j^{-1}(x_i) \qquad p = 1, 2, \ldots$$

Because of the nature of the inverse function, the merit value M will approach infinity on the constraint boundaries. The solution method using this penalty function proceeds as follows. First, a feasible design point in the allowable space is chosen inside the constrained set. Using any desired technique, the optimum is found for a preselected starting value of r_1. Since the starting point is inside the constrained area, the optimum found will also be within this area. Any trajectory of steepest descent leading from the internal design point cannot penetrate the boundary. Once the minimum is achieved, its location becomes the starting point for a new merit function in which the value of r_p is reduced. The optimization process is repeated for a sequence of decreasing r_p values:

$$r_1 > r_2 > r_3 > r_4 > \ldots > 0$$

As r_p approaches zero, the solution to the unconstrained optimization problem approaches the solution to the constrained problem.

Two simple penalty merit functions that have been used by Eason are

$$M(x_i) = F(x_i) + 10^{20} \sum_{j=1}^{j} A_j |G_j(x_i)|$$

and

$$M(x_i) = F(x_i) + W \sum_{j=1}^{j} A_j [G_j(X_i)]^2$$

where

$$A_j = \begin{cases} 1 & \text{if } G_j(x_i) < 0 \quad \text{(constraints violated)} \\ 0 & \text{if } G_j(x_i) \geq 0 \quad \text{(constraints not violated)} \end{cases}$$

Mischke suggests the use of a function of the form

$$M(x_i) = F(x_i) + \sum_{j=1}^{N} b[x_i(\text{good}) - x_i]^2$$

where

$$b = \begin{cases} 1 & \text{if } G_j(x_i) < 0 \quad \text{(constraints violated)} \\ 0 & \text{if } G_j(x_i) \geq 0 \quad \text{(constraints not violated)} \end{cases}$$

and x_i (good) represents a vector corresponding to a design that does not violate the constraints. The penalty function part of this merit surface is a smooth, unimodal, second-order hypersurface having its apex placed at a merit value of zero. The apex is located at the known point x_i (good). This penalty surface provides gradients that direct all stray search excursions back into the feasible design space.

2-3 CLASSIFICATION OF OPTIMIZATION ALGORITHMS

Numerical optimization algorithms can be classified according to the number of design variables and further according to the nature of the design space. Figure 2-5 depicts the different classes of optimization techniques to be covered in this book. In Chapter 3 we discuss the domain of one-dimensional search methods. In the remaining chapters we discuss various methods for multidimensional search.

2-4 PRACTICAL CONSIDERATIONS IN THE SETUP OF OPTIMIZATION PROBLEMS FOR THE SMALL COMPUTER

Although it is impossible to state universal rules that will guide the user in establishing the best form for a particular optimization problem, a few basic guidelines do exist. These are as follows:

1. *Consider the number of design variables.* If the user can reduce the number of design variables by means of an equality constraint, this procedure is almost always desirable. Indeed, there are situations where the type of algorithm to be used will depend strongly on the number of design variables. If the problem has only one design variable, the algorithms available are far superior to those available for multidimensional optimization. Also, as the number of design variables decreases, the amount of time required to achieve a solution also decreases. This can be especially important when using the small computer since run times for optimization problems can be significant.

2. *Consider the form of the merit function.* Whenever possible, the merit function should be expressed in the simplest possible form. The reasons for this are obvious. The merit function is often called as many as several hundred times during the implementation of a given optimization algorithm. Thus if the merit function is easily evaluated (formulated in such a way that it does not require extensive time to compute), considerable time saving will result.

3. *Consider the relative sizes of the design variables.* Whenever possible, it is wise to scale an optimization problem so that the design variables are of the same relative magnitude. Most optimization algorithms work best when this type of scaling is employed.

4. *Consider the nature of trade-offs.* Trade-offs occur whenever there are multiple objectives to be satisfied. For most optimization algorithms, only a single

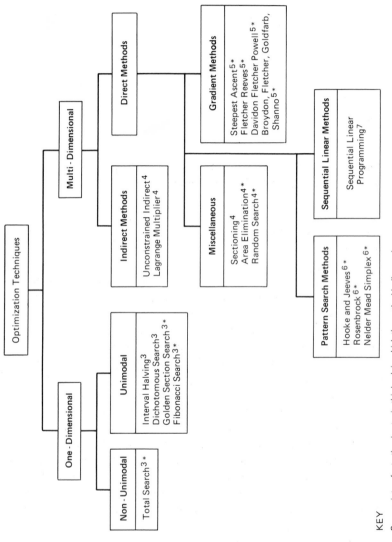

Figure 2-5 The family of optimization techniques.

KEY

Superscripts refer to the chapter in this book in which the method is discussed.
* means that software is provided in this book for this particular algorithm.

merit function can be utilized. If multiple merit measures must be optimized simultaneously (such as minimum costs, minimum weight, and maximum strength), a single merit function will have to be assembled, which is the linear sum of the various merit factors each multiplied by a weighting coefficient. The size of the weighting coefficient will indicate the relative importance of each individual element in the composite merit function.

REFERENCES

1. Carroll, C. W., "An Operations Research Approach to the Economic Optimization of a Kraft Pulping Process," doctoral dissertation, The Institute of Paper Chemistry, Appleton, Wis., 1959.

2. Converse, A. O., *Optimization,* Holt, Rinehart and Winston, New York, 1970.

3. Eason, E. E., and Fenton, R. G., Testing and Evaluation of Numerical Optimization Methods for Engineering Design, *ASME Paper 73-DET-17,* 1974.

4. Fiacco, A. V., and McCormick, G. P., "Computational Algorithm for the Sequential Unconstrained Minimization Technique for Nonlinear Programming," *Management Science.* Vol. 10 1964, pp. 601–617.

5. Fiacco, A. V., and McCormick, G. P., *Nonlinear Sequential Unconstrained Minimization Techniques,* John Wiley & Sons, Inc., New York, 1968.

6. Fiacco, A. V., and McCormick, G. P., "Computational Algorithm for the Sequential Unconstrained Minimization Technique for Nonlinear Programming, A Primal-Dual Method," *Management Science.* Vol. 10, 1964, pp. 360–366.

7. Fox, R. L., *Optimization Methods for Engineering Design,* Addison-Wesley Publishing Co., Inc., Reading, Mass., 1971.

8. Ketter, R. L., and Prawel, S. P., *Modern Methods of Engineering Computation,* McGraw-Hill Book Company, New York, 1969.

9. Mischke, C. R., *An Introduction to Computer-Aided Design,* Prentice-Hall, Inc., Englewood Cliffs, N.J., 1968.

10. Shoup, T. E., *A Practical Guide to Computer Methods for Engineers,* Prentice-Hall, Inc., Englewood Cliffs, N.J., 1979.

11. Zahradnik, R. L., *Theory and Techniques of Optimization for Practicing Engineers,* Barnes & Noble Books, New York, 1971.

3

ONE-DIMENSIONAL OPTIMIZATION

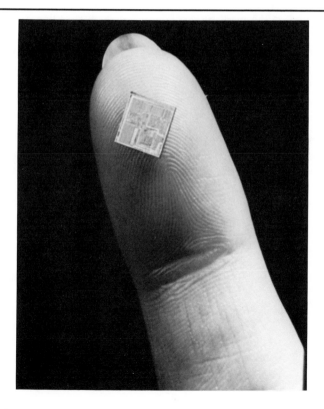

Although small in size, the potential uses for the microprocessor chip are nearly unlimited
(Courtesy of Intel.)

The search for extremes can be compared to the process of finding the deepest point in a lake by means of a series of soundings taken from a boat using a string and a weight. For each sounding, new information is gained. If each new depth is larger than previous trials, useful information is gained. If, on the other hand, the new depth is less than previous measurements, the new value is of no use and represents wasted effort. In search methods it is desired to reach the extremum as quickly as possible with a minimum of wasted effort. In this chapter one-dimensional search techniques are presented to facilitate the process of finding an optimum merit value for a one-dimensional merit function.

3-1 UNIMODAL MERIT FUNCTIONS

In this chapter it is assumed that the merit functions being investigated are *unimodal*. This means that they have only a single peak in the interval of interest. Thus as we make merit evaluations by slowly increasing the design variable, each successive merit value is progressively larger until we reach the peak. Once past the peak, each successive merit value is progressively less than the previous value. Actually, this limitation on the merit surface is not as restrictive as one might think since many problems in engineering exhibit this type of ''single-peak'' behavior.

The problem of one-dimensional optimization can be viewed within the following framework. A design variable x must have values between some lower and upper bound so that

$$a \leq x \leq b$$

As the problem begins, nothing is known about the merit function except that it must be unimodal. The *interval of uncertainty* will be defined to be the interval in which the optimum must lie. At the start of the optimization process, this interval is the total length from a to b, as shown in Figure 3-1.

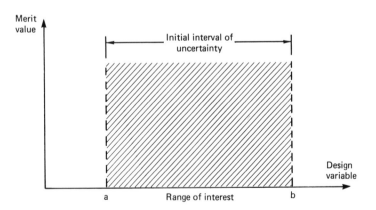

Figure 3-1 The interval of uncertainty.

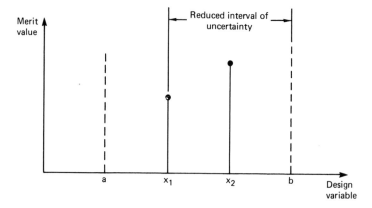

Figure 3-2   Reduction of the interval of uncertainty by means of two merit evaluations.

If two evaluations are made somewhere within the allowable range (say, at x_1 and x_2 to get M_1 and M_2), the interval of uncertainty is reduced, as shown in Figure 3-2. Several techniques for systematically reducing this interval exist. These will now be explored.

3-2 TOTAL SEARCH

Perhaps the most obvious technique to use in reducing the interval of uncertainty for a one-dimensional unimodal problem is to divide the total interval from a to b into a lattice of equally spaced functional evaluation, shown in Figure 3-3. As a result of this process the interval of uncertainty reduced to a value equal to two lattice spac-

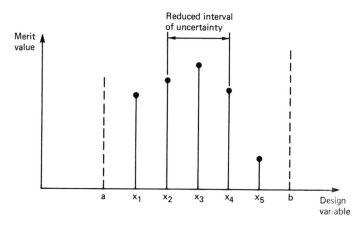

Figure 3-3   The total search technique.

ings. It is customary to speak of the fractional reduction of the interval of uncertainty f. In this case, for N lattice evaluations, the spacing will be

$$\frac{1}{N + 1}$$

and the fractional reduction of the interval of uncertainty will be

$$f = \frac{2}{N + 1}$$

To achieve a reduction of $f = 0.01$ will require $N = 199$ evaluations, and to achieve a reduction of $f = 0.001$ will require $N = 1999$. Clearly, for this method the efficiency of effort becomes poor as the desired size of the interval of uncertainty gets small. As a logical alternative to achieve $f = 0.01$ it would be better to expend 19 evaluations to get $f = 0.1$ and then expend an additional 19 evaluations on this new, smaller interval of uncertainty to achieve $f = 0.01$ in 38 evaluations rather than 199. Thus, with a bit of care, the efficiency of the search can be improved. Before proceeding with a presentation of an improved method, let us look at an example application involving the total search algorithm.

EXAMPLE 3-1

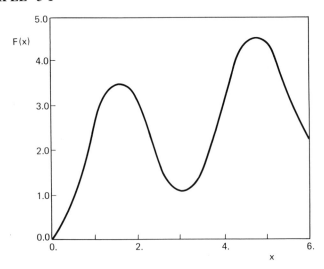

Suppose that it is desired to find the optimum of the function

$$F(x) = \frac{x}{3} + 3 \sin^2 x$$

on the range $0 \leq x \leq 6$. This relatively simple merit function is plotted above and is clearly nonunimodal. For this reason it helps us to illustrate the use of the total search technique.

Suppose further that it is desired to reduce the interval of uncertainty by a factor of 0.01 on the interval. A BASIC program containing a subroutine that performs the total search task follows.

```
100   GOTO 3000
110 :
120 :
500   REM ********************
510   REM * MERIT SUBROUTINE  *
520   REM ********************
530 :
540 F = X / 3 + 3 *  SIN (X) ^ 2
550   RETURN
560 :
570 :
2000   REM ********************
2010   REM * THIS SUBROUTINE   *
2020   REM * APPLIES THE TOTAL *
2030   REM * SEARCH ALGORITHM  *
2040   REM * TO FIND THE MAXI- *
2050   REM * MUM OF A SINGLE   *
2060   REM * DIMENSION MERIT   *
2070   REM * FUNCTION.         *
2080   REM *                   *
2090   REM * PARAMETERS:       *
2100   REM *                   *
2110   REM *                   *
2120   REM *   A,B - LEFT AND  *
2130   REM *         RIGHT BOUND*
2140   REM *         ON STARTING*
2150   REM *         INTERVAL OF*
2160   REM *         UNCERTAINTY*
2170   REM *                   *
2180   REM *   R   - FRACTIONAL *
2190   REM *         REDUCTION  *
2200   REM *         DESIRED OF *
2210   REM *         INTERVAL OF*
2220   REM *         UNCERTAINTY*
2230   REM *                   *
2300   REM * XL,XR- NEW LEFT   *
2310   REM *         AND RIGHT  *
2320   REM *         BOUNDS ON  *
2330   REM *         THE FINAL  *
2340   REM *         INTERVAL OF*
2350   REM *         UNCERTAINTY*
2360   REM *         FOUND.     *
2370   REM *                   *
2380   REM * XM   - OPTIMUM     *
2390   REM *         DESIGN     *
2400   REM *         VALUE FOUND*
2410   REM *                   *
2420   REM * FM   - OPTIMUM     *
2430   REM *         MERIT VALUE*
2440   REM *         FOUND      *
2450   REM *                   *
2460   REM * 500  - SUBROUTINE *
2470   REM *         TO COMPUTE *
2480   REM *         MERIT VALUE*
2500   REM ********************
```

```
2510 :
2540 L = B - A:D = .5 * R * L
2550 INC = 2 / R
2560 X = A: GOSUB 500
2570 FM = F:XM = A
2580  FOR I = 1 TO INC
2590 X = A + I * D
2600  GOSUB 500
2610  IF (F < FM) THEN  GOTO 2700
2620 FM = F:XM = X
2700  NEXT I
2710 :
2780  IF (XM = A) GOTO 2800
2790  IF (XM = B) GOTO 2830
2795 XL = XM - D:XR = XM + D
2798  RETURN
2800 XL = A:XR = A + D
2810  RETURN
2830 XL = B - D:XR = B
2840  RETURN
2850 :
2860 :
2870 :
3000  REM ********************
3010  REM * DRIVER PROGRAM    *
3020  REM ********************
3030 :
3040 A = 0:B = 6:R = .01
3050 :
3060 GOSUB 2000
3070 :
3080 PRINT "OPTIMUM VALUE FOUND=";FM
3090 PRINT "OPTIMUM ORDINATE   =";XM
3100 PRINT "---------------------------
--"
3110 PRINT "FINAL NEW UPPER AND LOWER"
3120 PRINT "BOUNDS ON THE INTERVAL OF"
3130 PRINT "UNCERTAINTY ARE:"
3140 PRINT "XL=";XL
3150 PRINT "XR=";XR
3160 :
3170 END
```

The careful observer will note that this program actually consists of three parts. These are a driver program (lines 3000–3170), a subroutine that implements the total search algorithm (lines 2000–2870), and a merit function routine (lines 500–570). These program portions have been arranged so that the most often used routines appear first in the listing. This arrangement will often save run time since a BASIC compiler will start searching for a subroutine from the first program line whenever a GOSUB command is given.

The output of this program is as follows:

```
OPTIMUM VALUE FOUND= 4.580055
OPTIMUM ORDINATE   = 4.77
--------------------------------
FINAL NEW UPPER AND LOWER
BOUNDS ON THE INTERVAL OF
UNCERTAINTY ARE:
XL= 4.74
XR= 4.8
```

This output required 9 seconds to complete on an IBM PC. A test run of this program for a value of R = 0.001 shows that the run time increases by a factor of 10 when the expected accuracy of the result is increased by a factor of 10. This experimental test result is easily explained since the higher accuracy requires 10 times more merit evaluations. This result should convince the reader that the total search algorithm is not an efficient approach if high accuracy is re- quired or if the merit function evaluation subroutine consumes large amounts of time in order to get a single merit value. One positive attribute of the subroutine in this example is that it does give the user a value of XL and XR to use as new values for A and B if the user decides to continue the search process after completing a cycle for a given choice of R.

3-3 INTERVAL HALVING

If we apply the logic described at the end of Section 3-2 but allow the number of evaluations in a given subsearch to be a variable, even more efficiency may be ob- tained. For N evaluations accomplished on I telescoping subsets, the final degree of reduction for the interval of uncertainty will be

$$f = \left(\frac{2}{N + 1} \right)^I$$

The total number of merit evaluations J expended in this search will be $J = N \times I$. It is desired to find the optimum N to minimize J for a given value of f. Using $I = J/N$, we can solve for J using the expression for interval of uncertainty. This expres- sion will be

$$J = \frac{N \ln (1/f)}{\ln [(N + 1)/2]}$$

If the value of J is plotted as shown in Figure 3-4, the minimum value is observed to lie somewhere close to 3. Since the number of evaluations must always be an inte- ger, the value N = 3 will be used as the optimum. for this choice $f = \frac{1}{2}$ for each subsearch. Since the interval of uncertainty is reduced by one-half, this technique is called *interval halving*.

Figure 3-5 illustrates how the initial three evaluations reduce the interval of uncertainty by one-half. Notice, however, that on this new interval, the central merit value is already known. Thus to complete the next subsearch, only two addi- tional evaluations are required rather than three. This saving continues as the algo-

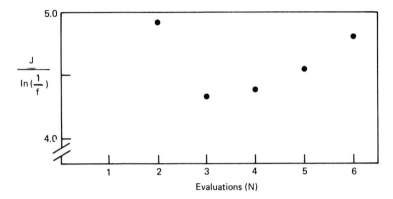

Figure 3-4 Locating the optimum number of merit evaluations for the interval halving method.

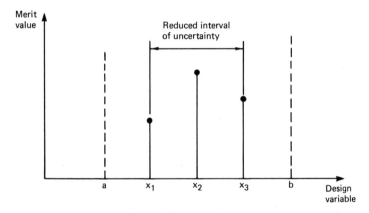

Figure 3-5 The first step in interval halving.

rithm proceeds. In general, the fractional reduction in uncertainty for N evaluations will be

$$f = \frac{1}{2^{\frac{(N-1)}{2}}} \qquad \text{for } N \geq 3$$

3-4 DICHOTOMOUS SEARCH

Throughout what has been discussed so far it has been required that the merit evaluations be equally spaced on the interval. If this restriction is lifted, it is possible to achieve a greater efficiency in the search. It has been shown that two merit evaluations on an interval will provide a reduction in the interval of uncertainty. Suppose that these trials are to be placed at special locations arranged so as to achieve the smallest interval of uncertainty as a result of their spacing. Figure 3-6 illustrates the

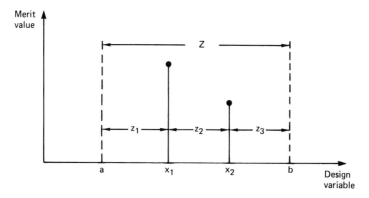

Figure 3-6   Notation used in the dichotomous search derivation.

notation used for this scheme. If the merit value of x_1 is greater than that of x_2, the new interval of uncertainty will be

$$Z_1 = z_1 + z_2$$

If the opposite is true, the new interval of uncertainty will be

$$Z_2 = z_2 + z_3$$

It is desired to minimize both Z_1 and Z_2 subject to the equality constraint $z_1 + z_2 + z_3 = Z$ and the regional constraints

$$0 < z_1$$
$$0 < z_2$$
$$0 < z_3$$

The equality constraint can be used to eliminate z_2 to get

$$Z - z_3 = \text{minimum}$$
$$Z - z_1 = \text{minimum}$$

Since Z is fixed, the larger z_3 and z_1 become, the smaller these equations will be. Thus the optimum will be

$$z_1 = z_3 = 0.5Z$$

But this would give $z_2 = 0$. Since this result violates one of the regional constraints, z_2 must be chosen to be some very small, nonzero value ϵ. A value $\epsilon/2$ will be subtracted from z_1 and z_3 to achieve compatability. For this choice the interval of uncertainty will be reduced to

$$f = \frac{1}{2} + \frac{\epsilon}{2}$$

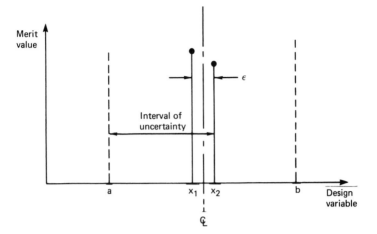

Figure 3-7 The dichotomous search.

for the first pair of closely spaced evaluations, as shown in Figure 3-7. In the limit as $\epsilon \rightarrow 0$ this uncertainty approaches $f = \frac{1}{2}$. This dichotomous search would then proceed just as we did with the interval halving method. Notice, however, that the same degree of reduction of the interval of uncertainty has been achieved with one less merit evaluation.

3-5 GOLDEN SECTION SEARCH

For any three merit evaluations on an interval of uncertainty, two will be useful for later evaluations, while one will not provide information of further use. It is the purpose of the *golden section search* to use a nonuniform spacing of merit evaluations arranged so that every single evaluation provides new and useful information. The basis for this scheme is as follows. In dividing an interval of uncertainty into two unequal parts, the ratio of the larger of the two segments to the total length of the interval should be the same as the ratio of the smaller to the larger segment. Thus if one considers an interval of uncertainty as shown in Figure 3-8 consisting of a length Z composed of two segments z_1 and z_2, the golden section theory requires that

$$\frac{z_1}{Z} = \frac{z_2}{z_1}$$

and

$$z_1 + z_2 = Z$$

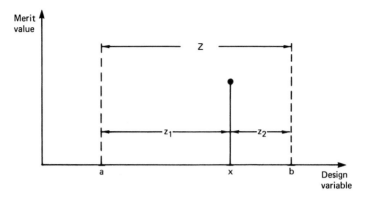

Figure 3-8   Notation used in the golden section search derivation.

The first equation gives

$$z_1^2 = Z z_2$$

Substituting for Z from the second equation and dividing through by z_1^2 gives

$$1 = \left(\frac{z_1}{z_2}\right)^2 + \left(\frac{z_1}{z_2}\right)$$

This quadratic can be solved for the ratio z_1/z_2. The positive root will be

$$\frac{z_1}{z_2} = 0.618033989$$

If the two evaluations on an interval are placed with this fractional spacing from either end, the result is as shown in Figure 3-9. Using these two evaluations, the interval of uncertainty will be reduced to a length of 0.618 times the previous interval of uncertainty. Although at this stage the method is not as good as interval halving by the dichotomous search, the true advantage is revealed as the new interval is further divided. Here it becomes obvious, by virtue of the golden ratio, that one of the two internal evaluations necessary for the next step is already available. Thus only one additional "golden-spaced" evaluation is required to reduce the uncertainty by another 0.618 fraction.

The golden section search surpasses the efficiency of the dichotomous search for $n > 2$ since every single evaluation provides an additional reduction of 0.618 for the interval of uncertainty. For N evaluations the final size for the fractional reduction in the interval of uncertainty will be

$$f = (0.618033989)^{N - 1}$$

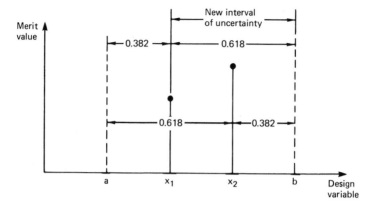

Figure 3-9 The golden section search.

The golden section search points out an interesting principle. To achieve the greatest reduction in subsequent intervals of uncertainty, the merit evaluations should be placed symmetrically about the centerline of the interval of uncertainty. When this is done and each new merit evaluation provides an additional reduction in the interval of uncertainty, the following mathematical expression applies:

$$Z_{I-2} = Z_{I-1} + Z_I \qquad 1 < I < N$$

where Z_I represents the length of the interval of uncertainty after the trial evaluation. It should be noted that other symmetrical techniques besides the golden section search can and do exist that satisfy this relationship. Before proceeding to look at these, let us consider an example of the use of the golden section search algorithm.

EXAMPLE 3-2

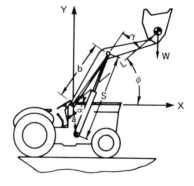

Modern loading machines frequently use a slider-crank mechanism with an oscillating slide made of a hydraulic cylinder to implement the deflection of their arm. An example of such a device is shown. The design process involves the selection of a geometric configuration that has a reasonable value for the maximum load in the hydraulic cylinder over the entire range of motion. Suppose that a particular design has been suggested with a boom length of 3.0 meters and a maximum load of 1500 kg. Find the maximum load seen by the hydraulic cylinder on the allowable range of motion

$$\phi_{min} = -20 \text{ deg} = -0.3490658504 \text{ rad}$$
$$\phi_{max} = 80 \text{ deg} = 1.39623402 \text{ rad}$$

if a particular design geometry has $C = 0.5850$ and $D = 1.2691$ with the value of $\beta = 1.2231$. The relationship between the cylinder load and the geometry can be found from the free-body diagram.

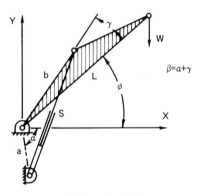

$$K_1 = 2 \, C \, D$$
$$K_2 = C^2 + D^2$$

The relationship for the cylinder load will be

$$T(\phi) = \frac{LW\sqrt{K_2 - K_1 \cos(\beta + \phi)} \cos \phi}{CD \sin(\beta + \phi)}$$

Thus the problem to be solved is to find the maximum of $T(\phi)$ on the design range

$$\phi_{min} \le \phi \le \phi_{max}$$

A BASIC computer program that implements this solution follows.

```
100   GOTO 3000
110   :
500   REM ********************
510   REM * MERIT SUBROUTINE *
520   REM * TO FIND THE LOAD *
530   REM * ON A HYDRAULIC   *
```

```
540   REM * CYLINDER IF GIVEN*
550   REM * THE GEOMETRIC     *
560   REM * CONFIGURATION AND*
570   REM * THE LOAD LIFTED. *
580   REM ********************
590 :
600 LL = 3:WW = 1500
610 BETA = 1.2231
620 C = .585:D = 1.2691
630 F = LL * WW * COS (X) * SQR(C ^ 2
+ D ^ 2 - 2 * C * D * COS (BETA + X)) /
(C * D * SIN (BETA + X))
640   RETURN
650 :
2000   REM ********************
2010   REM * THIS SUBROUTINE   *
2020   REM * APPLIES THE GOLDEN*
2030   REM * SEARCH ALGORITHM  *
2040   REM * TO FIND THE MAXI- *
2050   REM * MUM OF A SINGLE   *
2060   REM * DIMENSION MERIT   *
2070   REM * FUNCTION.         *
2080   REM *                   *
2090   REM * PARAMETERS:       *
2100   REM *                   *
2110   REM *                   *
2120   REM *   A,B - LEFT AND  *
2130   REM *         RIGHT BOUND*
2140   REM *         ON STARTING*
2150   REM *         INTERVAL OF*
2160   REM *         UNCERTAINTY*
2170   REM *                   *
2180   REM *   R   - FRACTIONAL *
2190   REM *         REDUCTION  *
2200   REM *         DESIRED OF *
2210   REM *         INTERVAL OF*
2220   REM *         UNCERTAINTY*
2230   REM *                   *
2240   REM * RR   - ACTUAL     *
2250   REM *         REDUCTION  *
2260   REM *         OF THE     *
2270   REM *         INTERVAL OF*
2280   REM *         UNCERTAINTY*
2290   REM *                   *
2300   REM * XL,XR- NEW LEFT    *
2310   REM *         AND RIGHT  *
2320   REM *         BOUNDS ON  *
2330   REM *         THE FINAL  *
2340   REM *         INTERVAL OF*
2350   REM *         UNCERTAINTY*
2360   REM *         FOUND.     *
2370   REM *                   *
2380   REM * XM   - OPTIMUM    *
2390   REM *         DESIGN     *
2400   REM *         VALUE FOUND*
2410   REM *                   *
2420   REM * FM   - OPTIMUM    *
2430   REM *         MERIT VALUE*
2440   REM *         FOUND      *
2450   REM *                   *
2460   REM * 500  - SUBROUTINE *
2470   REM *         TO COMPUTE *
2480   REM *         MERIT VALUE*
2490   REM *         F (X).     *
2500   REM ********************
```

```
2510 :
2520  REM  GOLDEN SEARCH
2530 XL = A:XR = B
2540 L = B - A
2550 LM = L * R
2560 X1 = A + .381966011# * L
2570 X2 = A + .618033989# * L
2580 X = X1: GOSUB 500:F1 = F
2590 X = X2: GOSUB 500:F2 = F
2600  IF F1 < F2 THEN  GOTO 2680
2610 XR = X2
2620 L = XR - XL
2630  IF L < LM GOTO 2750
2640 X2 = X1:F2 = F1
2650 X1 = XL + .381966011# * L
2660 X = X1: GOSUB 500:F1 = F
2670  GOTO 2600
2680 XL = X1
2690 L = XR - XL
2700  IF L < LM GOTO 2750
2710 X1 = X2:F1 = F2
2720 X2 = XL + .618033989# * L
2730 X = X2: GOSUB 500:F2 = F
2740  GOTO 2600
2750 :
2760  REM  DESIRED REDUCTION
2770 RR = L / (B - A)
2780 FM = F2:XM = X2
2790  IF F2 > F1 THEN  GOTO 2810
2800 FM = F1:XM = X1
2810  RETURN
2820 :
3000  REM *********************
3010  REM * THIS DRIVER PROGRAM*
3020  REM * FINDS THE MAXIMUM  *
3030  REM * LOAD IN A HYDRAULIC*
3040  REM * CYLINDER OF A LIFT *
3050  REM * MECHANISM OVER ITS *
3060. REM * RANGE OF MOTION.   *
3070  REM *********************
3080 :
3090  REM  SET RANGE OF MOTION
3100 A =  - .3490658504#
3110 B = 1.39623402#
3120 :
3130  REM  SET REDUCTION IN
3140  REM  INTERVAL OF
3150  REM  UNCERTAINTY DESIRED.
3160 R = .001
3170 :
3180  REM  USE GOLDEN SEARCH
3190  GOSUB 2000
3200 :
3210  PRINT "MAX LOAD=";FM;" (KG)"
3220  PRINT "LOCATION IS AT ";XM;" (RAD)
"
3230  END
```

The output of this program is as follows:

```
MAX LOAD= 7966.267  (KG)
LOCATION IS AT  .3890199  (RAD)
```

This optimum required 3 seconds to complete on an IBM PC.

3-6 FIBONACCI SEARCH

Although the golden section search works quite well, it is obviously not an optimum for a given number of evaluations. For example, if the designer knows in advance that only two function evaluations can be used, it is likely that the designer will utilize a dichotomous pair to reduce the interval of uncertainty to 0.5 rather than accept the 0.618 possible through the golden section search. If the designer is allowed to adjust the spacing of the merit evaluations as the search progresses, the combined advantages of the symmetrical method mentioned previously can be combined with the use of a final dichotomous pair to achieve an optimum search algorithm. Suppose that Z_n is the length of the interval of uncertainty after the nth trial. The symmetry requirement for the evaluation will be

$$Z_{i-2} = Z_{i-1} + Z_i \qquad 1 < i < n$$

and the requirement that the last evaluation be a dichotomous spacing of ϵ is

$$Z_{n-1} = 2Z_n - \epsilon$$

From these two relationships it is possible to work backward to determine the required size for any intermediate interval of uncertainty and thus also determine the placement positions for the merit evaluations. For example,

$$\begin{aligned} Z_{n-3} &= Z_{n-2} + Z_{n-1} \\ &= (Z_{n-1} + Z_n) + Z_{n-1} \\ &= 5Z_n - 2\epsilon \end{aligned}$$

and

$$Z_{n-4} = 8Z_n - 3\epsilon$$

The general relationship for any interval of uncertainty will be

$$Z_{n-k} = F_{k+1}(Z_n) - F_{k-1}$$

where the F_j coefficients are called *Fibonacci numbers* and are defined in terms of

$$\begin{aligned} F_0 &= 1 \\ F_1 &= 1 \\ F_k &= F_{k-1} + F_{k-2} \qquad \text{for } k = 2, 3, \ldots \end{aligned}$$

Table 3-1 presents some of the Fibonacci numbers. In general, the size of the final interval of uncertainty can be expressed as

$$Z_n = \frac{1}{F_n} + \frac{F_{n-2}}{F_n}\epsilon$$

TABLE 3-1 THE FIBONACCI NUMBERS

k	F_k	k	F_k
0	1		
1	1	11	144
2	2	12	233
3	3	13	377
4	5	14	610
5	8	15	987
6	13	16	1597
7	21	17	2584
8	34	18	4181
9	55	19	6765
10	89	20	10946

In the limit as $\epsilon \to 0$ a lower bound can be determined on the size of the smallest interval of uncertainty that can be achieved for a given number of evaluations.

The procedure for applying the Fibonacci method is first to decide how many merit evaluations (n) can be used. Then information about the length of the interval of uncertainty can be used to plan the merit evaluation spacing. Since $Z_1 = Z_0 = 1$ this process must start with two evaluations placed Z_2 distance from opposite ends of the initial interval. In this case

$$Z_2 = \frac{F_{n-1}}{F_n} + \frac{(-1)^n}{F_n}\epsilon$$

where ϵ represents the smallest distance by which two evaluations may be separated and still be distinguished from one another. The next step is to locate an additional evaluation Z_3 units from the end of Z_2. The appropriate interval is retained and the process is repeated until the nth evaluation has been carried out.

The Fibonacci search algorithm can best be illustrated by working an example problem.

EXAMPLE 3-3

Suppose that it is desired to work the problem posed in Example 3-2 by the Fibonacci search algorithm. A BASIC program containing a subroutine to implement this algorithm follows.

```
100  GOTO 3000
110  :
500  REM ********************
510  REM * MERIT SUBROUTINE *
520  REM * TO FIND THE LOAD *
530  REM * ON A HYDRAULIC   *
```

```
540   REM * CYLINDER IF GIVEN*
550   REM * THE GEOMETRIC    *
560   REM * CONFIGURATION AND*
570   REM * THE LOAD LIFTED. *
580   REM ********************
590   :
600 LL = 3:WW = 1500
610 BETA = 1.2231
620 C = .585:D = 1.2691
630 F = LL * WW *  COS (X) *  SQR(C ^ 2
 + D ^ 2 - 2 * C * D * COS (BETA + X)) /
(C * D * SIN (BETA + X))
640   RETURN
650   :
2000    REM ********************
2010    REM * THIS SUBROUTINE   *
2020    REM * USES THE FIBONACCI*
2030    REM * SEARCH ALGORITHM  *
2040    REM * TO FIND THE MAXI- *
2050    REM * MUM OF A SINGLE   *
2060    REM * DIMENSION MERIT   *
2070    REM * FUNCTION.         *
2080    REM *                   *
2090    REM * PARAMETERS:       *
2100    REM *                   *
2110    REM *  A,B - LEFT AND   *
2120    REM *        RIGHT BOUND*
2130    REM *        ON STARTING*
2140    REM *        INTERVAL OF*
2150    REM *        UNCERTAINTY*
2160    REM *                   *
2170    REM *  R   - FRACTIONAL *
2180    REM *        REDUCTION  *
2190    REM *        DESIRED OF *
2200    REM *        INTERVAL OF*
2210    REM *        UNCERTAINTY*
2220    REM *                   *
2230    REM * RR   - ACTUAL     *
2240    REM *        REDUCTION  *
2250    REM *        OF THE     *
2260    REM *        INTERVAL OF*
2270    REM *        UNCERTAINTY*
2280    REM *                   *
2290    REM * XL,XR- NEW LEFT   *
2300    REM *        AND RIGHT  *
2310    REM *        BOUNDS ON  *
2320    REM *        THE FINAL  *
2330    REM *        INTERVAL OF*
2340    REM *        UNCERTAINTY*
2350    REM *        FOUND.     *
2360    REM *                   *
2370    REM * XM   - OPTIMUM    *
2380    REM *        DESIGN     *
2390    REM *        VALUE FOUND*
2400    REM *                   *
2410    REM * FM   - OPTIMUM    *
2420    REM *        MERIT VALUE*
2430    REM *        FOUND      *
2440    REM *                   *
2450    REM * 500  - SUBROUTINE *
2460    REM *        TO COMPUTE *
2470    REM *        MERIT VALUE*
2480    REM *        F(X).      *
2490    REM ********************
2500    :
2510 FI(0) = 1:FI(1) = 1
```

```
2520 XL = A:XR = B:NN = 2
2530 L = B - A:E = L * R * .001
2540 :
2550  REM  LOAD MAX FIBONACCI
2560  REM  NUMBER NEEDED
2570  FOR N = 2 TO 20
2580 FI(N) = FI(N - 1) + FI(N - 2)
2590  IF (1 / FI(N) < R) THEN  GOTO 2620
2600  NEXT N
2610 :
2620 Z = L * FI(N - 1) / FI(N)
2630 X2 = A + Z:X1 = B - Z
2640 X = X1: GOSUB 500:F1 = F
2650 X = X2: GOSUB 500:F2 = F
2660 :
2670  IF F1 < F2 GOTO 2820
2680 :
2690  REM  HERE F1 IS LARGEST
2700 XR = X2:NN = NN + 1
2710 X2 = X1:F2 = F1
2720  IF N = NN GOTO 2780
2730  IF N < NN GOTO 2940
2740 Z = L * FI(N - NN) / FI(N)
2750 X1 = XL + Z
2760 X = X1: GOSUB 500:F1 = F
2770  GOTO 2670
2780 X1 = X2 - E
2790 X = X1: GOSUB 500:F1 = F
2800  GOTO 2670
2810 :
2820  REM  HERE F2 IS LARGEST
2830 XL = X1:NN = NN + 1
2840 X1 = X2:F1 = F2
2850  IF N = NN GOTO 2910
2860  IF N < NN GOTO 2940
2870 Z = L * FI(N - NN) / FI(N)
2880 X2 = XR - Z
2890 X = X2: GOSUB 500:F2 = F
2900  GOTO 2670
2910 X2=X1+E
2920 X=X2: GOSUB 500: F2=F
2930  GOTO 2670
2940 :
2950  REM  SOLUTION FOUND
2960 FM = F2:XM = X2
2970 RR = 1 / FI(N): RETURN
2980 :
2990 :
3000  REM *********************
3010  REM * THIS DRIVER PROGRAM*
3020  REM * FINDS THE MAXIMUM  *
3030  REM * LOAD IN A HYDRAULIC*
3040  REM * CYLINDER OF A LIFT *
3050  REM * MECHANISM OVER ITS *
3060  REM * RANGE OF MOTION.   *
3070  REM *********************
3080 :
3090  DIM FI(20)
3100 :
3110  REM  SET RANGE OF MOTION
3120 A =  - .3490658504#
3130 B = 1.39623402#
3140 :
3150  REM  SET REDUCTION IN
3160  REM  INTERVAL OF
3170  REM  UNCERTAINTY DESIRED.
```

```
3180  R = .001
3190  :
3200   REM   USE FIBONACCI SEARCH
3210   GOSUB 2000
3220  :
3230   PRINT "MAX LOAD=";FM;" (KG)"
3240   PRINT "LOCATION IS AT ";XM;" (RAD)
"
3350 END
```

The output of this program is as follows:

```
MAX LOAD= 7966.268  (KG)
LOCATION IS AT  .3886157  (RAD)
```

This output required 9 seconds to complete on an IBM PC. It is interesting to compare this output with that for the golden search algorithm. The Fibonacci program required 16 merit evaluations to achieve the desired degree of accuracy. In general, the Fibonacci algorithm requires fewer merit evaluations than the golden search to achieve the same degree of reduction of the final interval of uncertainty. The price paid for the use of fewer merit evaluations is the time spent to compute the Fibonacci numbers that are required to implement the solution process. For problems where the merit evaluation was quite involved, the Fibonacci algorithm can often save considerable time. For problems with a simple merit function (such as this one), the golden search algorithm is probably best to use.

The careful observer will note that the merit subroutine is placed early in the program listing. This was done for an important reason. Most basic interpreters found on personal computers scan for a subroutine whenever a GOSUB command is found. The scan process starts with the first program lines and proceeds to the end. If the routine that is called for is found early in the scan, the program is not delayed as much as it would be if the routine were at the end. This leads to a faster program if the subroutine is used many times. Thus the general rule to observe in applying the personal computer is to place first the routine used most often and follow this with the routines used next most often, in order of their frequency of use.

Both the Fibonacci search routine and the golden search routine are written in such a way that the user could restart them for an additional reduction in the interval of uncertainty since the output of one run provides the new lower and upper bounds for the reduced interval of uncertainty.

3-7 COMPARISON OF ONE-DIMENSIONAL SEARCH TECHNIQUES

The five search techniques for the personal computer described previously can best be compared in terms of two measures of effectiveness. These measures are the overall algorithm efficiency and the programming utility.

A common measure of *efficiency* for an algorithm is the number of functional evaluations required to achieve a desired degree of accuracy. Table 3-2 indicates that the Fibonacci search method is best by this measure and the total search ranks lowest. Because of their efficiency, the designer will usually find that the Fibonacci and the golden section search methods are the best to try first on a unimodal optimization problem in one dimension. The choice as to which of these to choose will depend on the nature of the merit function. If the merit function does not require much time to calculate, the two algorithms appear to function equally well. If the merit function is mathematically complex and requires considerable computer time to evaluate, the Fibonacci algorithm is a better choice.

TABLE 3-2 COMPARISON OF THE ONE-DIMENSIONAL SEARCH METHODS: FRACTIONAL REDUCTION OF THE INTERVAL OF UNCERTAINTY[a]

Number of Merit Evaluations, n	Total Search f	Interval Halving f	Dichotomous Search f	Golden Section Search f	Fibonacci Search f
1	1.0	1.0	1.0	1.0	1.0
2	0.667	—	0.500	0.618	0.500
3	0.500	0.500	—	0.382	0.333
4	0.400	—	0.250	0.236	0.200
5	0.333	0.250	—	0.146	0.125
6	0.286	—	0.125	0.090	0.077
7	0.250	0.125	—	0.056	0.048
8	0.222	—	0.0625	0.0345	0.0294
9	0.200	0.625	—	0.0213	0.0182
10	0.182	—	0.0312	0.0132	0.0112
11	0.167	0.0312	—	0.00813	0.00694
12	0.154	—	0.0156	0.00502	0.00429
13	0.143	0.0156	—	0.00311	0.00265
14	0.133	—	0.00781	0.00192	0.00164
15	0.125	0.00781	—	0.00119	0.00101
16	0.118	—	0.00391	0.000733	0.000626
17	0.111	0.00391	—	0.000453	0.000387
18	0.105	—	0.00195	0.000280	0.000239
19	0.100	0.00195	—	0.000173	0.000148
20	0.095	—	0.000976	0.000107	0.0000913

[a]All numbers are rounded to three significant digits.

The *utility* of an algorithm describes the ease with which it can be applied and the applicability it has to a variety of problem situations. In terms of utility, the Fibonacci search method is probably the most difficult to apply since it requires a separate calculation to determine the location of the merit evaluation for each new step. This, of course, is the price one must pay to achieve the high efficiency of this algorithm. In terms of utility, the inefficient total search does have one virtue. This is its ability to succeed on nonunimodal merit functions if they are reasonably well behaved. Frequently, the designer does not know whether the merit function is unimodal. When this situation exists, the designer should try several different types of algorithms to see if they each converge to the same optimum. This situation points out an important principle in optimization. No single algorithm can solve every problem. The designer should attempt to become proficient at using a variety of algorithms in order to increase his or her success rate when considering difficult optimization problems.

Although the topic of one-dimensional optimization is actually a subset of the more general topic, multivariable optimization, one-dimensional search needs occur so frequently in design and their methodology is so unique that they have been discussed separately. In Chapter 4 the concept of optimization is expanded to the more challenging multidimensional search methods.

PROBLEMS

3-1 It is desired to design a new type of storage container consisting of an open, circular cone as shown. If the container is to hold exactly a volume of 1 m^3, what will be the dimensional size of the design having minimum surface area?

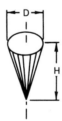

3-2 If the container in Problem 3-1 has a circular lid, find the dimensional size of the design having minimum surface area.

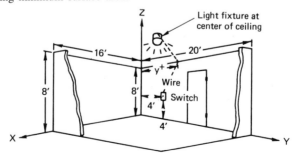

3-3 A new, cast concrete, modular concept has been proposed for the construction of hotel rooms. The prefabricated rectangular modules are to be "prewired" before being placed in position. Since a large number of these units will be made to identical specifications, it is essential that the cost of materials be minimized. As part of the design task you have been asked to specify the location of the electrical wire from a wall switch to the ceiling fixture. The design is to use a minimum length of wire and must always be in the wall or in the ceiling. Specify the path that the wire must travel by locating the dimension y* and specify the total nominal length of wire required for this task.

3-4 A rectangular highway sign of dimensions w and h is to be manufactured from sheet metal of uniform thickness. The sign must have a printed area of 1.5 m², a 20-cm margin at the bottom, and 10-cm margins along the other three sides, as indicated. If it is desired to use minimum material, find the best choice for the dimensions w and h.

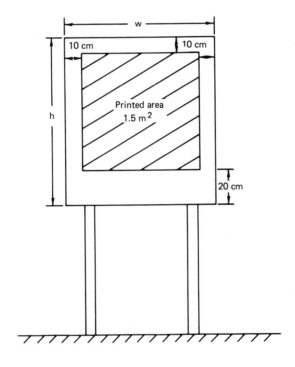

3-5 How would your answer to Problem 3-4 be different if the following constraints were added?

$$1.4 \leq w \leq 2.0 \text{ m}$$

3-6 Calculate the maximum range (D) and the corresponding firing angle (θ) for the cannon shown.

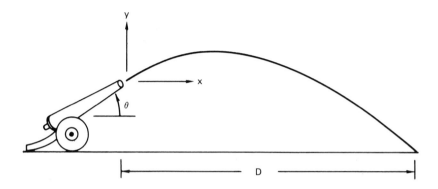

The initial velocity of the projectile is

$$V_0 = 800 \text{ m/s}$$

The mass of the projectile is

$$m = 50 \text{ kg}$$

Assume that the firing takes place on a flat plane,

$$y(0) = y_{\text{final}}$$

and that the drag on the projectile is

$$F_d = CV^2$$

where $C = 0.015$ N-s^2/m.

3-7 Under what conditions would an engineering optimization problem have no solution? Can you think of an example of this situation?

3-8 The discharge flow rate for steady, fully developed laminar flow of an incompressible fluid through the annual space between two concentric tubes of circular cross section is

$$Q = K \left[r_o^4 - r_i^4 - \frac{r_o^2 - r_i^2}{\ln (r_o/r_i)} \right]$$

where K is a constant that depends on the pressure drop per unit length, the fluid density, and the viscosity. It is desired to design the geometry (choose r_i and r_o) of the two concentric cylinders so as to maximize the flow rate through a cross-sectional area of 10 cm^2 subject to the restriction that

$$2.0 \leq r_i \leq r_o \leq 10 \text{ cm}$$

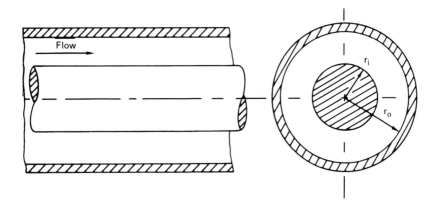

3-9 One creative way to find a root of a general polynomial is by minimizing a merit function that approaches zero magnitude when a root is approached. Try this method using a one-dimensional search method for the polynomial

$$x^5 + 21x^4 + 158x^3 + 502x^2 + 609x + 245 = 0$$

3-10 A cylindrical container that is closed at one end only is to be designed with a volume of 10 m³. If the diameter of the cylinder is D and the height of the cylinder is h, design a container that uses minimum outside surface area.

REFERENCES

1. Converse, A. O., *Optimization,* Holt, Rinehart and Winston, New York, 1970.
2. Fox, R. L., *Optimization Methods for Engineering Design,* Addison-Wesley Publishing Co., Inc., Reading, Mass., 1971.
3. Ketter, R. L. and Prawel, S. P., *Modern Methods of Engineering Computation,* McGraw-Hill Book Company, New York, 1969.
4. Kuester, J. L. and Mize, J. H., *Optimization Techniques with FORTRAN,* McGraw-Hill Book Company, New York, 1973.
5. Mischke, C. R., *An Introduction to Computer-Aided Design,* Prentice-Hall, Inc., Englewood Cliffs, N.J., 1968.
6. Shoup, T. E., *A Practical Guide to Computer Methods for Engineers,* Prentice-Hall, Inc., Englewood Cliffs, N.J., 1979.
7. Zahradnik, R. L., *Theory and Techniques of Optimization for Practicing Engineers,* Barnes & Noble Books, New York, 1971.

4

INTRODUCTION TO MULTIDIMENSIONAL SEARCH METHODS

The video-tracing system increases the speed and accuracy of tracing images into a microcomputer CAD system. (Courtesy of BrighterImages.)

At first thought the designer may be tempted to believe that the difference between multidimensional search techniques and single-dimensional search techniques is only one of increased effort, and if the designer were willing to spend a bit more time in the calculation process he or she could extend single-variable methods to N-dimensional methods. Unfortunately, this is not true since the nature of multidimensional space is considerably different from one-dimensional space. For one thing, as the number of dimensions increases, the likelihood that a merit function will be unimodal decreases. In addition, the size of multidimensional space is overwhelming. The degree of effort required to achieve a given degree of reduction of the interval of uncertainty increases by the exponent of the dimensionality of the space. For instance, if in one-dimensional space 19 evaluations are needed to achieve $f = 0.1$, then 361 evaluations will be required to achieve the same accuracy in two dimensions, 6859 in three dimensions, 130,321 in four dimensions, and 2,476,099 in five dimensions. Since it is not uncommon to have five or more design variables in a general optimization problem, the seriousness of multidimensionality becomes painfully obvious.

Traditionally, optimization methods in multidimensional space are classified in terms of two broad categories called direct methods and indirect methods. *Direct methods* utilize a comparison of functional evaluation; *indirect methods* employ the mathematical principles of maximization or minimization. Direct methods attempt to establish a strategy to "zero in" on the optimum; indirect methods attempt to satisfy the conditions of the problem without examining nonoptimal points. In the following chapters we look at various methods currently in use for multidimensional optimization. The present chapter focuses on indirect methods and on the fundamental direct approaches to simplifying the nature of the multidimensional design space. The more sophisticated direct algorithms are covered in the following chapters.

4-1 INDIRECT OPTIMIZATION

No treatment of multidimensional optimization techniques would be complete without a discussion of the calculus of stationary points. For a multidimensional function to have a maximum, a minimum, or a saddle point, it is necessary that all first derivatives with respect to each of the n independent variables be zero. Thus for the function

$$M(x) = F(x_1, x_2, \ldots, x_n)$$

a stationary point will satisfy

$$\frac{\partial F}{\partial x_1} = 0, \quad \frac{\partial F}{\partial x_2} = 0, \quad \ldots, \quad \frac{\partial F}{\partial x_n} = 0$$

To determine whether a stationary point is a minimum, a maximum, or a saddle point, it is necessary to examine the second derivatives of the function. A convenient way to describe the nature of the second derivatives is by means of the Hessian matrix, which is of the form

$$\text{Hessian} = \begin{bmatrix} \dfrac{\partial^2 F}{\partial x_1^2} & \dfrac{\partial^2 F}{\partial x_1\,\partial x_2} & \cdots & \dfrac{\partial^2 F}{\partial x_1\,\partial x_n} \\[2ex] \dfrac{\partial^2 F}{\partial x_1\,\partial x_2} & \cdots\cdots\cdots\cdots\cdots\cdots\cdots\cdots \\[2ex] \cdot & & \\ \cdot & & \\ \cdot & & \\[2ex] \dfrac{\partial^2 F}{\partial x_n\,\partial x_1} & \cdots\cdots\cdots\cdots\cdots\cdots & \dfrac{\partial^2 F}{\partial x^2_n} \end{bmatrix}$$

A necessary and sufficient condition for a stationary point to be a local minimum is that its Hessian matrix be positive definite. This means that all its eigenvalues will be positive. A necessary and sufficient condition for a stationary point to be a local maximum is that its Hessian matrix be negative definite. This means that all its eigenvalues will be negative. One way to mechanize this information is shown in Figure 4-1. First the system of equations corresponding to the n first

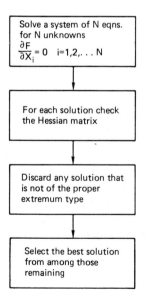

Figure 4-1 The indirect method of optimization.

partial derivatives is found. This system must be solved for all possible sets of design values that satisfy the equations. If these equations are linear, the problem is straightforward since only one solution set will exist. If the system is nonlinear, as is most often the case, there may be many solution sets. Once the solution sets are isolated, the designer must discard all the solution sets that are not of the desired extremum type. This requires a check of the eigenvalues of the Hessian matrix of second partial derivatives evaluated at each of the solution design points. Once the solution sets have been reduced to a final group, the designer must check to see which of the group has the most desirable merit value. This one will be declared the optimum. Although the foregoing technique does seem mathematically straightforward, it is, in reality, not extremely practical since the optimum in many design situations will occur at a boundary rather than at a stationary point. The technique does point out the need for methods to extract eigenvalues and for methods to solve systems of nonlinear algebraic equations.

One interesting extension of the technique of stationary points is the method of Lagrange multipliers. This technique has the advantage of allowing equality constraints of the form

$$Q_1(x_1, x_2, \ldots, x_n) = 0$$
$$\cdot$$
$$\cdot$$
$$\cdot$$
$$Q_j(x_1, x_2, \ldots, x_n) = 0$$

to be satisfied in the optimization process. To facilitate the solution of this problem, a new merit function must be formed that is a linear combination of the old merit function and each of the constraint equations multiplied by a unique constant. This new merit function will be

$$M(x_i, \lambda_i) = F(x_i) + \lambda_1 Q_1 + \lambda_2 Q_2 + \ldots + \lambda_j Q_j$$

The λ_j values are called *Lagrange multipliers* and are said to be treated as additional unknowns to be determined in the solution process. Thus the system used to locate stationary points consists of $j + n$ equations and $j + n$ unknowns. If each of the constraints is satisfied, the additional λ_j terms each contribute nothing to the new merit function. In this case the optimization of M is equivalent to the optimization of F. It should be noted that in the equations to be solved for the stationary point, the partial derivatives of the new merit function with respect to the unknown Lagrangian multipliers revert to the constraint equations.

4-2 SECTIONING METHOD

A logical extension of the one-dimensional search methodology discussed previously would be to alter one independent design variable at a time until the merit value ceases to improve, then to do the same to each individual variable in sequence. Once the last variable has been treated, the designer can repeat the process starting with the first variable to see if additional improvement can be achieved. This sectioning or "one-at-a-time" search method will not always reach the optimum. Figure 4-2a illustrates a case of contours that are well suited for this technique. The characteristic of this surface is that its contours approximate either cir-

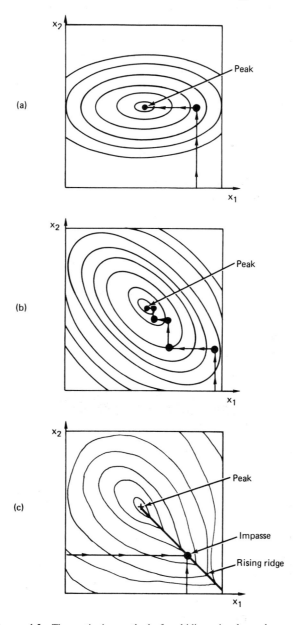

Figure 4-2 The sectioning method of multidimensional search.

cles or ellipses with their major or minor axes parallel to the coordinate axes. If these contour axes are tilted relative to the coordinate axes as shown in Figure 4-2b, the algorithm efficiency decreases because the method is forced to use many more evaluations to achieve an optimum answer. The method fails completely whenever it encounters a ridge where the contours come to a point, as shown in Figure 4-2c. Since ridges of this type can frequently occur in engineering design problems, this technique should not be used unless the designer can be sure that a particular problem is free of this pitfall. Nevertheless, the sectioning procedure does find application as a starting procedure for more complex search techniques. The advantage of this method is that it allows the designer to use one of the one-dimensional search algorithms (such as the golden section search) that are well understood and are noted for their efficiency of effort.

4-3 AREA ELIMINATION

After seeing how effective one-dimensional techniques are at reducing the interval or area of uncertainty, one might wish to consider the existence of similar methodology in multidimensional space. One of the most obvious area elimination schemes is called the *contour tangent method* because it makes use of a tangent to the contours of the merit function. This technique can best be visualized in terms of the top view of a two-dimensional design merit surface, as shown in Figure 4-3. Here an arbitrarily selected, feasible point in design space has been found to lie on a contour somewhere below the optimum peak. In the plane of the contour, the tangent to the contour can be located. This tangent is not difficult to find since it must lie in the plane of the contour and must also be perpendicular to the local gradient of the merit surface at the design point. If the merit surface is well behaved and is strongly unimodal, the contour tangent will divide the feasible design space into regions that have high promise or low promise in so far as the extremum is concerned. Using

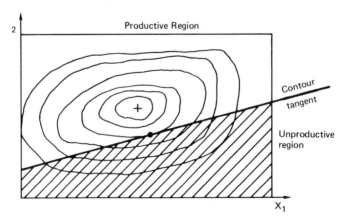

Figure 4-3 The area elimination method of search.

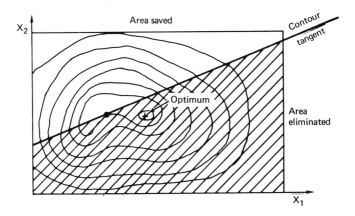

Figure 4-4   The area elimination method of search.

this technique and several well-placed merit observations, the designer can substantially reduce the search domain. There is a difficulty associated with the use of this algorithm, however. If a contour line is concave rather than convex, as shown in Figure 4-4, the possibility of eliminating the region containing the extremum does exist. In addition, the region of uncertainty remaining after several eliminations may be awkward to manipulate by other algorithms.

One type of area elimination technique that does give manageable results is the *grid search* of Mischke. In this technique the reduced area of uncertainty will be a hypercube (i.e., the multidimensionalization of a square or cube) of predictable size. For this reason it is one of the few multidimensional techniques with measureable effectiveness. Visualization in terms of a design space with two design variables will aid in the understanding of this search technique. The original region of uncertainty is mapped onto a unit square, cube, or hypercube (depending on the dimension of the space) so that the search is normalized to a region of unit dimension on a side. Through this hypercube is drawn a grid of symmetrically paired, orthogonal planes parallel to the design variable axes. The intersections of planes will generate lines that will in turn intersect to give points known as nodes, as shown in Figure 4-5. The merit values at each intersection and at the center of the cube are evaluated. This involves $2^m + 1$ total functional evaluations for m design variables. A note is made of the largest merit value found, and this location becomes the center of a smaller hypercube for further investigation. The process continues until the desired degree of reduction in the interval of uncertainty is achieved. If the fractional reduction in allowable domain along any design variable axis is r, then the linear fractional reduction for b hypercubes will be

$$F = r^b$$

and the total function evaluations for process will be

$$n = b(2^m) + 1$$

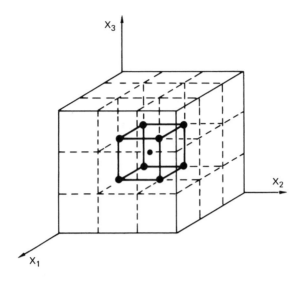

Figure 4-5   The grid search technique.

The added 1.0 on the formula is for the central merit position. As the process proceeds, the best merit ordinate found is used as the center of a smaller design space. For the degree of reduction of the design space from stage to stage, Mischke recommends that

$$\frac{2}{3} < r < 1$$

be used and comments that a *star pattern* is more efficient than grid node points for problems with three or more design variables. For the star pattern, functional evaluations are made at a central ordinate value and at adjacent values corresponding to a single step in each ordinate direction. For this procedure the total number of functional evaluations will be

$$n = b(2^{m-1}) + m + 1$$

Figure 4-6 provides a description of the procedure used for the star-pattern search area elimination method.

One of the important advantages of the area elimination method is that the user can know in advance how many function evaluations (and thus the amount of computational effort) will be needed in order to reduce the interval of uncertainty by a desired amount. Before proceeding with another algorithm, let us consider an example application for the area elimination method.

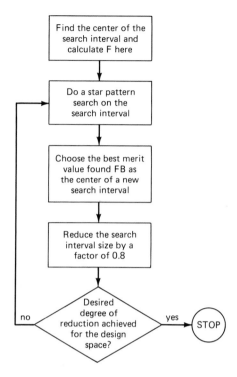

Figure 4-6   The star-pattern search area elimination algorithm.

EXAMPLE 4-1

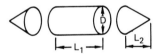

It is desired to design a new type of aircraft fuel storage pod, as shown. The symmetrical pod is to be fabricated from three sections of metal sheet material and will consist of a cylindrical center with conical ends. The three pieces, once fabricated, will be welded together. The design specifications for this pod call for a volume of 1.0 m^3. It is desired to minimize the amount of material used in the design for reasons of weight and cost. What dimensions L_1, L_2 and D would you recommend for this design?

For this design problem, the quantity to be minimized will be the total surface area of the pod:

$$A = L_1 \pi D + 2(\pi \frac{D}{2} \sqrt{\frac{D^2}{4} + L_2^2})$$

The need for one of the three design variables L_1, L_2, and D can be eliminated using the equality constraint

$$\text{volume} = 1.0 = \pi D^2 (\frac{L_2}{6} + \frac{L_1}{4})$$

Thus

$$D = \sqrt{\frac{1}{\pi[(L_2/6) + (L_1/4)]}}$$

Thus when problem to be solved is written in standard form the *design variables* are

$$L_1 \quad \text{and} \quad L_2$$

and the *merit function* to be minimized is

$$A = \sqrt{\frac{\pi}{(L_2/6) + (L_1/4)}} \left[L_1 + \sqrt{\frac{1}{4} \frac{1}{\pi[(L_2/6) + (L_1/4)]}} + L_2^2 \right]$$

As now formulated this problem is an unconstrained optimization problem for two design variables. Thus the star-pattern search area elimination method can be used. To implement this method, it will be necessary to establish upper and lower bounds for the size of the design variables. As a first selection we will use

$$0 \le L_1 \le 1$$
$$0 \le L_2 \le 1$$

A BASIC program containing a subroutine that performs this search follows. This routine reduces the final interval of uncertainty by a factor of $R = 0.001$ over the starting interval. The subroutine implements the algorithm presented in Figure 4-6 and makes use of a star pattern with legs the size of one-fourth of the design variable range for each stage of the process. The subroutine uses a shrinkage factor of 0.8 for each reduced interval of the design space. These two choices for subroutine parameters could, of course, be adjusted to a variety of numerical values; however, the user should be careful to pick a set that will not exclude any areas of the design space.

```
100 GOTO 3000
101 :
500 REM ********************
510 REM * MERIT SUBROUTINE *
520 REM ********************
530 :
540 B = X(2)/6+X(1)/4
550 F = SQR(PI/B)*(X(1) + SQR(1/(4*PI*B)
 + X(2)^2))
560 RETURN
570 :
580 :
1000   REM ********************
1010   REM * THIS SUBROUTINE    *
1020   REM * APPLIES AN AREA     *
1030   REM * ELIMINATION METHOD*
1040   REM * TO FIND AN UNCON-   *
1050   REM * STRAINED MINIMUM    *
1060   REM * OF A MERIT          *
1070   REM * FUNCTION.   THE     *
1080   REM * USER MUST SUPPLY    *
1090   REM * UPPER AND LOWER     *
1100   REM * BOUNDS FOR THE      *
1110   REM * DESIGN VARIABLES.   *
1120   REM *                     *
1130   REM * PARAMETERS:         *
1140   REM *                     *
1150   REM *   NV -   THE NUMBER *
1160   REM *          OF DESIGN  *
1170   REM *          VARIABLES. *
1180   REM *                     *
1190   REM *   R  -   FRACTIONAL *
1200   REM *          REDUCTION  *
1210   REM *          DESIRED OF *
1220   REM *          THE DESIGN *
1230   REM *          SPACE      *
1240   REM *                     *
1250   REM * XR,XL- UPPER AND    *
1260   REM *          LOWER BOUND*
1270   REM *          VECTORS FOR*
1280   REM *          THE DESIGN *
1290   REM *          VARIABLES. *
1300   REM *                     *
1310   REM * 500   - SUBROUTINE  *
1320   REM *          TO EVALUATE*
1330   REM *          THE MERIT  *
1340   REM *          VALUE "F"  *
1350   REM *          USING A    *
1360   REM *          DESIGN VEC-*
1370   REM *          TOR "X(I)" *
1380   REM *                     *
1390   REM * FB   - ON RETURN    *
1400   REM *          THE BEST   *
1410   REM *          MERIT VALUE*
1420   REM *          FOUND.     *
1430   REM *                     *
1440   REM * XB(I)- THE DESIGN   *
1450   REM *          VECTOR FOR *
1460   REM *          FB.        *
1470   REM ********************
1480 :
1490   DIM XC(NV),DX(NV)
```

```
1500 :
1510  REM  FIND CENTER POINT
1520  REM  AND SET STEP SIZE
1530  FOR I = 1 TO NV
1540 X(I) = XL(I) + .5 * (XR(I) -  XL(I)
1550 XB(I) = X(I)
1560 XC(I) = X(I)
1570 DX(I) = (XR(I) - XL(I)) / 4
1580  NEXT I
1590  GOSUB 500:FB = F
1600 S = DX(1)
1610 :
1620 :
1630  REM  EXPLORE IN STAR
1640  REM  PATTERN
1650  FOR I = 1 TO NV
1660 X(I) = XC(I) + DX(I)
1670  GOSUB 500
1680  IF (F > FB) GOTO 1730
1690 FB = F
1700  FOR II = 1 TO NV
1710 XB(II) = X(II)
1720  NEXT II
1730 X(I) = XC(I)
1740  NEXT I
1750  FOR I = 1 TO NV
1760 X(I) = XC(I) - DX(I)
1770  GOSUB 500
1780  IF (F > FB) GOTO 1830
1790 FB = F
1800  FOR II = 1 TO NV
1810 XB(II) = X(II)
1820  NEXT II
1830 X(I) = XC(I)
1840  NEXT I
1850 :
1860  REM  SET NEW CENTER
1870  REM  POINT AT XB(I)
1880  REM  AND DECREASE STEP
1890  FOR I = 1 TO NV
1900 XC(I) = XB(I)
1910 DX(I) = .8 * DX(I)
1920 X(I) = XB(I)
1930  NEXT I
1940 :
1950  REM  IS REDUCTION SMALL
1960  REM  ENOUGH
1970  IF DX(1) / S > R THEN  GOTO 1630
1980 RETURN
2100 :
3000  REM *********************
3010  REM *THIS DRIVER PROGRAM*
3020  REM *PERFORMS THE DESIGN*
3030  REM *OF A SYMMETRICAL   *
3040  REM *CYLINDRICAL TANK OF*
3050  REM *VOLUME OF 1 M^3    *
3060  REM *WITH MINIMUM       *
3070  REM *SURFACE AREA.      *
3080  REM *********************
3090 :
3100 R = .001:NV = 2
3110 PI = 3.1415926#
3120  DIM X(NV),XL(NV),XR(NV)
3130  DIM XB(NV)
3140 :
3150 XL(1) = 0:XR(1) = 1
```

```
3160 XL(2) = 0:XR(2) = 1
3170 :
3180  GOSUB 1000
3190 :
3200  PRINT "-------------------- "
3210  PRINT "AREA=";FB; TAB( 17); "M^2"
3220  PRINT "L(1)=";XB(1); TAB( 17);"M"
3230  PRINT "L(2)=";XB(2); TAB( 17);"M"
3240  GOSUB 500
3250 D =  SQR (1 / (PI * B))
3260  PRINT "D   =";D; TAB( 17);"M"
3270  PRINT "--------------------"
3280 :
3290  END
```

The output of this program is as follows:

```
--------------------
AREA= 5.019215  M^2
L(1)= .5348248  M
L(2)= .5345933  M
D   = 1.19526   M
--------------------
```

This output required 9 seconds to complete on an IBM PC. The form of the output suggests that the two design values L_1 and L_2 are probably equal for the optimum case. Since the solution values are well away from the established boundaries, the solution represents a true minimum in the stationary-point sense. If, on the other hand, the solution had tended to be along one of the boundaries, it would be good to try a wider initial range for the design boundaries. The price paid for a larger starting domain is that the final answers found will have a correspondingly lower final domain of uncertainty for a given choice of R. The nature of the algorithm does not allow the search to consider any value outside the range specified for the initial domain of uncertainty. This subroutine can be adapted for use for a wide variety of engineering design problems.

4-4 RANDOM SEARCH

Earlier in this chapter the vastness of multidimensional space was discussed in terms of how many evaluations would be necessary to achieve $f = 0.1$ by a lattice approach. It was found that the number increases by the exponent of the dimension of the space. An interesting and creative alternative to circumvent this difficulty, proposed by Brooks (1958), is based on a random selection process. Suppose that the design space is viewed as a unit cube or hypercube and is divided into cubical cells by means of 10 equally spaced divisions along each design variable axis. If $n = 2$ there will be 100 cells, if $n = 3$ there will be 1000 cells, and in general there will be 10^n cells for a space having n design variables. The probability that one cell chosen at random will be among the best 10 percent is 0.1 since we would be looking for

TABLE 4-1

	Probability			
f	0.80	0.90	0.95	0.99
0.1	16	22	29	44
0.05	32	45	59	90
0.01	161	230	299	459
0.005	322	460	598	919

the one best in 10 for $n = 1$, one of the 10 best in 100 for $N = 2$, and so on. The probability of missing one of the best 10 percent will be 0.9. If two cells are selected at random, the probability of missing becomes $(0.9)^2$ or 0.81. In general, the probability of finding at least one cell in the best fraction f when N random selections are made will be

$$P = 1 - (1 - f)^n$$

Table 4-1 provides a listing of the number of random selections needed to achieve a specific probability for a desired best fraction. From this table it can be seen that 44 random selections will have a 99 percent probability of achieving $f = 0.1$. This represents an attractive alternative to the 2,476,099 evaluations required to guarantee $f = 0.1$ by total enumeration with five design variables.

The random process has two attractive features. First, the method works well on any type of merit surface, whether unimodal or not. Second, the probability of success in n random selections does not depend on the dimension of the space being

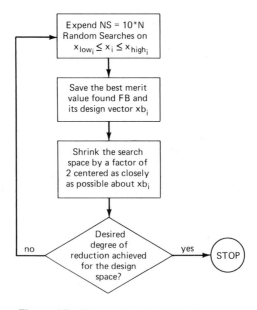

Figure 4-7 The random search algorithm.

considered. Although this technique does not move toward an optimum, it does provide a way to achieve a "good" starting point to be used by other search methods. For this reason it is often used in combination with another algorithm of a different type. The basic logic to the use of this algorithm is illustrated in Figure 4-7. The method is based on the use of groups of random searches applied to a design space that shrinks in size at the end of each search group. To illustrate the use of this algorithm, consider the following example application.

EXAMPLE 4-2

Suppose that it is desired to solve the problem posed in Example 4-1 by a random search method. A BASIC program containing a subroutine that performs this search follows. This routine reduces the final interval of uncertainty by a factor R = 0.001 over the starting interval:

$$0 < L_1 < 1$$
$$0 < L_2 < 1$$

The subroutine makes use of the algorithm presented in Figure 4-7.

```
100 GOTO 3000
101 :
500 REM *******************
510 REM * MERIT SUBROUTINE *
520 REM *******************
530 :
540 B = X(2)/6+X(1)/4
550 F = SQR(PI/B)*(X(1) + SQR(1/(4*PI*B)
    + X(2)^2))
560 RETURN
570 :
580 :
1000   REM ********************
1010   REM * THIS SUBROUTINE   *
1020   REM * APPLIES A RANDOM  *
1030   REM * SEARCH TO FIND AN *
1040   REM * UNCONSTRAINED     *
1050   REM * MINIMUM OF A MERIT*
1060   REM * FUNCTION.  THE    *
1070   REM * USER MUST SUPPLY  *
1080   REM * UPPER AND LOWER   *
1090   REM * BOUNDS FOR THE    *
1100   REM * DESIGN VARIABLES. *
1110   REM *                   *
1120   REM * PARAMETERS:       *
1130   REM *                   *
1140   REM * NV -  THE NUMBER  *
1150   REM *       OF DESIGN   *
1160   REM *       VARIABLES.  *
1170   REM *                   *
1180   REM * R  -  FRACTIONAL  *
1190   REM *       REDUCTION   *
1200   REM *       DESIRED OF  *
1210   REM *       THE DESIGN  *
1220   REM *       SPACE       *
```

```
1230  REM *                       *
1240  REM * XR,XL- UPPER AND  *
1250  REM *         LOWER BOUND*
1260  REM *         VECTORS FOR*
1270  REM *         THE DESIGN *
1280  REM *         VARIABLES. *
1290  REM *                       *
1300  REM * 500  - SUBROUTINE *
1310  REM *         TO EVALUATE*
1320  REM *         THE MERIT  *
1330  REM *         VALUE "F"  *
1340  REM *         USING A    *
1350  REM *         DESIGN VEC-*
1360  REM *         TOR "X(I)" *
1370  REM *                       *
1380  REM * FB  - ON RETURN   *
1390  REM *         THE BEST   *
1400  REM *         MERIT VALUE*
1410  REM *         FOUND.     *
1420  REM *                       *
1430  REM * XB(I) -THE DESIGN *
1440  REM *         VECTOR FOR *
1450  REM *         FB.        *
1460  REM ********************
1470  :
1480  DS = XR(NV) - XL(NV)
1490  NS = 10 * NV
1500  :
1510  REM  FIND A START
1520  FOR I = 1 TO NV
1530  X(I) = XL(I) + .5 * (XR(I) - XL(I))
1540  XB(I) = X(I)
1550  NEXT I
1560  GOSUB 500:FB = F
1570  :
1580  REM  EXPEND 10N RANDOM
1590  REM  SEARCHES TO FIND
1600  REM  THE BEST FB VALUE
1610  FOR J = 1 TO NS
1620  FOR I = 1 TO NV
1630  X(I)=XL(I)+RND(NV)*(XR(I)-XL(I))
1640  NEXT I
1650  GOSUB 500
1660  IF (F>FB) GOTO 1710
1670  FB=F
1680  FOR I=1 TO NV
1690  XB(I)=X(I)
1700  NEXT I
1710  NEXT J
1720  :
1730  REM  SHRINK THE DESIGN
1740  REM  SPACE BY A FACTOR
1750  REM  OF 2 ABOUT XB(I).
1760  :
1770  FOR I=1 TO NV
1780  D=XR(I)-XL(I)
1790  :
1800  REM IS THE BEST VALUE
1810  REM SUFFICIENTLY INSIDE
1820  REM THE DESIGN SPACE?
1830  :
1840  IF XB(I)+D/4>XR(I) THEN GOTO 1900
1850  IF XB(I)-D/4<XL(I) THEN GOTO 1940
1860  XL(I)=XB(I)-D/4
1870  XR(I)=XB(I)+D/4
1880  GOTO 1940
```

```
1890 :
1900 REM   TOO CLOSE TO XR
1910 XL(I)=XR(I)-D/2
1920 GOTO 1960
1930 :
1940 REM   TOO CLOSE TO XL
1950 XR(I)=XL(I)+D/2
1960 NEXT I
1970 :
1980 IF D/DS>R THEN GOTO 1580
1990 RETURN
2000 :
3000   REM *********************
3010   REM *THIS DRIVER PROGRAM*
3020   REM *PERFORMS THE DESIGN*
3030   REM *OF A SYMMETRICAL   *
3040   REM *CYLINDRICAL TANK OF*
3050   REM *VOLUME OF 1 M^3    *
3060   REM *WITH MINIMUM       *
3070   REM *SURFACE AREA.      *
3080   REM *********************
3090 :
3100 NR=VAL(RIGHT$(TIME$,2))
3110 RANDOMIZE(NR)
3120 :
3130 R = .001:NV = 2
3140 PI = 3.1415926#
3150 DIM X(NV),XL(NV),XR(NV)
3160 DIM XB(NV)
3170 :
3180 XL(1) = 0:XR(1) = 1
3190 XL(2) = 0:XR(2) = 1
3200 :
3210   GOSUB 1000
3220 :
3230   PRINT "-------------------- "
3240   PRINT "AREA=";FB; TAB( 17); "M^2"
3250   PRINT "L(1)=";XB(1); TAB( 17);"M"
3260   PRINT "L(2)=";XB(2); TAB( 17);"M"
3270   GOSUB 500
3280 D =   SQR (1 / (PI * B))
3290   PRINT "D   =";D; TAB( 17);"M"
3300   PRINT "--------------------"
3310 :
3320   END
```

The output of this program is as follows:

```
--------------------
AREA= 5.019215  M^2
L(1)= .5342233  M
L(2)= .5348519  M
D   = 1.195586  M
--------------------
```

This output required 19 seconds to complete on an IBM PC.

An interesting fact about the use of an algorithm based on random searching is that each time it is used, the output will be slightly different. For example, if the program is run again, the result will be as follows:

```
--------------------
AREA= 5.019215  M^2
L(1)= .5345649  M
L(2)= .5350132  M
D   = 1.195247  M
--------------------
```

This subroutine utilizes stages that use random searches of the design space. The number of random searches used is 10 times the number of design variables. The design space is reduced in size after each stage by a factor of 2. These two parameters could, of course, be adjusted to meet the needs of a particular design situation. There is a trade-off between the probability of finding a good answer and the time required to locate that answer. The random search subroutine is especially well suited to use in finding a good starting value for other types of search algorithms.

4-5 COMPARISON OF THE SECTIONING METHOD, THE AREA ELIMINATION METHOD, AND THE RANDOM METHOD

The three search techniques for the personal computer described in this chapter can best be compared in terms of the types of situations in which they work best.

The sectioning method has the considerable advantage that it allows the designer to make use of the golden search and Fibonacci search methods. These one-dimensional methods are extremely efficient and generally guarantee a solution for a minimum of effort in a one-dimensional space. There are multidimensional problems that can be handled efficiently by this approach. Unfortunately, as the independent variables become more interactive, the method fails to give rapid convergence. This "slow convergence" situation can usually be observed by monitoring the progress of the solution through PRINT statements inserted in the driver program. When the approach does not appear productive, it is wise to abandon the method in favor of one of the other, more efficient methods.

The area elimination method has the advantage of being the only multidimensional method that allows the user to know in advance exactly how many merit evaluations will be required to achieve a desired degree of reduction in the final interval of uncertainty. This, of course, can be a strong advantage. A user may choose an initial value for R that is relatively large in order to get an idea of the run time required on a particular personal computer. Using this initial information and the number of design variables, the user can predict quite accurately how much time will be required to achieve a much finer reduction of the interval of uncertainty. The primary disadvantage of the method is that it is based on the assumption of unimodality. If the user suspects the design space to be multimodal, he or she should try different starting points. The method requires the user to choose boundaries for the design variables, and the method will refuse to look for solutions outside this domain. This limitation can cause the user to overlook possible good solutions if the boundaries are chosen too conservatively.

The random method is an interesting approach to optimization because it does not require that the merit function be unimodal. Unfortunately, it carries no guarantee that the optimum solution found will be the best possible. For this reason, the random method is often used as a means to find reasonable starting values required by other, more complex methods.

PROBLEMS

4-1 A waste treatment settling tank is to be designed to hold 40,000 liters of liquid waste. The tank is made of 10-cm concrete with reinforcing. Design the tank so that a minimum amount of concrete is used.

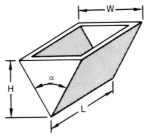

4-2 Solve Problem 4-1 if the container has a closed top.

4-3 A container manufacturer is designing an open-top container of sheet material. The sheet is to be cut, folded on the dashed lines, and welded along the four seams. If the outside surface area of the box cannot exceed 1.0 m and if no dimension (a, b, c) can be larger than three times any other dimension, what size container will have maximum volume?

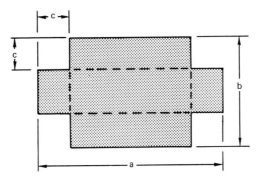

4-4 It is often convenient to be able to find the best-fit straight line for an arbitrary set of experimental data points. The best-fit straight line is expressed as

$$y = mx + b$$

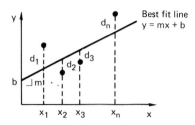

The accumulated error for k data points is expressed as

$$\text{sum} = \sum_{i=1}^{k} (d_i^2)^{j/2}$$

Write a computer program that will perform the minimization of the sum to find the optimum values of M and b for:
(a) $j = 1$ (absolute value of error)
(b) $j = 2$ (least-squares error)
(c) $j = 4$ (fourth power of error)

for the data

i	x_i	y_i
1	1.0	1.0
2	1.5	2.8
3	2.8	2.5
4	2.7	4.0
5	4.3	3.0

4-5 If Problem 4-4 is solved using the best fit of a second-order equation $y = a_1 x^2 + a_2 x + a_3$, find the optimum values for a_1, a_2, and a_3 for the three cases.

4-6 It is desired to choose the dimensions for the slider-crank mechanism shown. It is desired to do this to provide the best approximation (minimum error) for the function

$$x_{\text{desired}} = 1 + \cos^2 \theta \qquad \text{for } 0° \le \theta \le 90°$$

The actual input versus output behavior for the device is

$$x_{\text{actual}} = \cos(\theta + \theta_0) + \sqrt{\cos^2(\theta + \theta_0) - (A_1^2 - A_2^2)}$$

You may express the error as

$$\text{error} = \int_0^{90°} (x_{\text{desired}} - x_{\text{actual}}) \, d\theta$$

The particular design requires that

$$A_1 < A_2$$
$$0.1 < A_1 < 10$$
$$0.1 < A_2 < 10$$

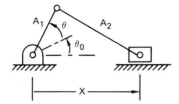

4-7 Solve the problem in Example 4-1 using the golden section search sectioning method. Is the method faster or slower than the area elimination method?

4-8 One creative way to solve a set of simultaneous, nonlinear algebraic equations is to treat them as if they were an optimization problem and try to minimize the sum of the absolute values of the residuals (the right-hand sides). Start with the point $(1, 1, 1, 1)$ and see if you can find a solution to the set of equations:

$$x_1 \quad + 2x_2 \quad + x_3^2 \quad + x_4^3 \quad - 30 = 0$$
$$x_1 \quad - 3x_2^2 \quad + x_3 \quad - x_4 + 10 = 0$$
$$2x_1 \quad - x_2x_3 + x_3x_4 \quad - 8 = 0$$
$$2x_1/x_2 \quad + x_3x_4^2 \quad - 49 = 0$$

on the range $0 \leq x_i \leq 5$ for $i = 1, 2, 3,$ and 4.

4-9 For the gear train shown, the speed ratio that relates the angular velocity of the last gear to that of the first gear depends on the number of teeth (N_i) on each gear in the train:

$$\text{speed ratio} = \frac{N_1N_3N_5}{N_2N_4N_6}$$

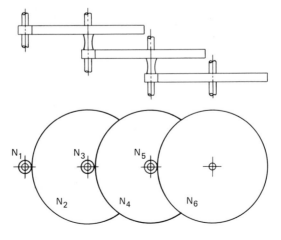

It is desired to design a gear train that will have a speed ratio that is as close to 1/10 as possible. For practical reasons the number of teeth on each gear must be restricted to

$$20 \leq N_i \leq 100 \qquad i = 1, 2, \ldots, 6$$

Find the best possible design by specifying the values of N_1, N_2, N_3, N_4, N_5, and N_6.

4-10 It is not uncommon to have an optimization problem in which a subsearch is required to find the merit value for another search process. Suppose, for example, that it is desired to find a curve from among the family of curves shown such that the value of α is found to give a curve with a peak at $M = \pi/12$. The family of curves is defined by

$$M(\phi, \alpha) = \phi - \frac{(\pi/2)(F(\phi, \alpha)}{F(\pi/2, \alpha))}$$

where $F(\phi, \alpha)$ is an elliptic integral of the first type:

$$F(\phi, \alpha) = \int_0^\phi (1 - \sin^2 \alpha \sin^2 \theta) \, d\theta$$

Prepare and run a computer program that will find the answer to this problem.

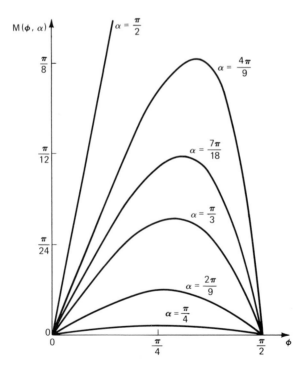

REFERENCES

1. Brooks, S. H., ''A Discussion of Random Methods for Seeking Maxima,'' *Operations Research,* Vol. 6, 1958 pp. 244–251.
2. Mischke, C. R., *An Introduction to Computer-Aided Design,* Prentice-Hall, Inc., Englewood Cliffs, N.J., 1968.
3. Shoup, T. E., *A Practical Guide to Computer Methods for Engineers,* Prentice-Hall, Inc., Englewood Cliffs, N.J., 1979.
4. Wilde, D. J., and Beightler, C. S., *Foundations of Optimization,* Prentice-Hall, Inc., Englewood Cliffs, N.J., 1967.

5

GRADIENT METHODS FOR MULTIDIMENSIONAL SEARCH

Recent software systems allow the personal computer to perform tasks once reserved for only the most advanced minicomputers. (Courtesy of Manufacturing and Consulting Services, Inc.)

A large number of multidimensional optimization algorithms depend in some way on gradient information. The basis for this fact can be seen in a simple illustration. Suppose that a mountain climber were blindfolded and told to climb to the top of a unimodal mountain. Even without the benefit of being able to see the peak, the climber could reach the top simply by remembering always to walk uphill. Although any rising path will eventually lead to the top, the path where the slope is steepest is the best, provided that the climber does not encounter a vertical cliff that he or she cannot scale. (The mathematical equivalent of a cliff would be a ridge caused by a constraint in the merit surface.) For now it will be assumed that the optimization problem is unconstrained. The constraints can, of course, be incorporated into the search scheme using penalty function methods. The optimization equivalent of the steepest path idea is known as the *method of steepest ascent* or the *method of steepest descent*. The gradient vector is perpendicular to a contour and can be used to locate a new design point. It should be noted that the gradient method, unlike the contour tangent method, can be used on any unimodal function, not just those that are strongly unimodal.

To understand the rationale of gradient-based methods, it is well to look into the nature of the gradient. Consider a system of independent unit vectors $\mathbf{e}_1$, $\mathbf{e}_2$, $\mathbf{e}_3$, . . . $\mathbf{e}_n$ that are aligned with the design variable axes $x_1, x_2, x_3, \ldots, x_n$. The gradient vector for a general merit function $F(x_1, x_2, x_3, \ldots x_n)$ will be of the form

$$\mathbf{gradient} = \frac{\partial F}{\partial x_1} \mathbf{e}_1 + \frac{\partial F}{\partial x_2} \mathbf{e}_2 + \cdots + \frac{\partial F}{\partial x_n} \mathbf{e}_n$$

where the partial derivatives are evaluated at the point being considered. This vector points in the upward or ascent direction and its negative points in the descent direction. The unit gradient vector is often written as

$$\mathbf{gradient} = v_1\mathbf{e}_1 + v_2\mathbf{e}_2 + v_3\mathbf{e}_3 + \cdots + v_n\mathbf{e}_n$$

where

$$v_i = \frac{\dfrac{\partial F}{\partial x_i}}{\left[\displaystyle\sum_{j=1}^{n} \left(\frac{\partial F}{\partial x_j}\right)^2\right]^{1/2}}$$

In some cases the nature of the merit function is well enough known to allow differentiation to calculate the gradient vector components. If the partial derivatives cannot be extracted in this way, they may be approximated by small local explorations to get

$$\frac{\partial F}{\partial x_i} = \frac{F(x_1, x_2, \ldots, x_i + \Delta x_i, \ldots, x_n) - F(x_1, x_2, \ldots, x_i, \ldots, x_n)}{\Delta x_i}$$

where Δx_i is a small exploration along the x_i direction. This formula is often referred to as the *secant approximation*. Once the gradient direction is known, it can be used in a variety of ways to implement a search strategy.

5-1 STEEPEST ASCENT BY STEPS

Some search methods move a fixed step up the gradient and reevaluate the function. If an improvement has been achieved, a new gradient is computed and the procedure is repeated, often with an increased step size. If no improvement or a negative improvement is found, the step size from the previous best point is decreased and the procedure is repeated. This process continues until no improvement can be achieved by decreasing the step size.

5-2 STEEPEST ASCENT BY ONE-DIMENSIONAL SEARCH

Some search methods use information about the gradient to conduct a one-dimensional search along the direction of the steepest ascent or descent using the relationship

$$x_i(\text{new}) = x_i(\text{old}) + S\,v_i$$

where S is the new one-dimensional parameter along the unit gradient v_i.
The unit gradient vector for the direction of steepest descent is:

$$v_i = \frac{-\dfrac{\partial F}{\partial x_i}}{\left[\displaystyle\sum_{j=1}^{n}\left(\frac{\partial F}{\partial x_j}\right)^2\right]^{1/2}} \qquad i = 1, 2, \ldots, n$$

Once the one-dimensional optimum along the gradient has been achieved, a new gradient is found and the process is repeated until no further improvement can be found. The primary advantage of this method is that the parameter S may be used as the independent variable for a Fibonacci search, and thus the method tends to be quite efficient in effort. One of the principal advantages of steepest gradient methods is their ability to steer away from saddle points on the merit surface, as indicated in Figure 5-1. It should be noted in this figure, however, that gradient techniques will find only a local optimum when applied to multimodal surfaces. For this reason, if the nature of the surface is not well known, several starting points should be tried to see if every start leads to the same optimum. The random optimization method described in Chapter 4 can often be used to find one or more feasible starting points in the design space in order to implement this strategy. Another difficulty

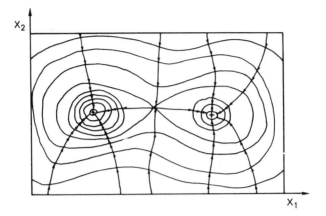

Figure 5-1 A bimodal merit surface.

that can hamper the efficiency of gradient methods occurs when the technique encounters a ridge. Since a ridge represents a discontinuity in the slope of a contour line, it tends to give false information on the proper direction to move. Thus the search technique may slow down and zigzag back and forth across the ridge, making progress toward the optimum quite slow. In some cases the severity of this performance on a ridge is so slow that the algorithm must be abandoned. In reality, however, a large number of merit surfaces associated with problems from engineering design have one or more ridges. These ridges often point toward the optimum. Thus the difficulty associated with ridges can sometimes be turned into an advantage. Whenever a ridge is encountered, the best direction to move is often along the ridge rather than in the direction of the local gradient. Several sophisticated search techniques utilize this approach and are referred to as accelerated climbing techniques. These methods are presented later in this chapter, but first let us consider an example application of the method of steepest descent.

EXAMPLE 5-1

Although it is rather simple mathematically, the Rosenbrock merit function

$$y = 100(x_2 - x_1^2)^2 + (1 - x_1)^2$$

contains a curved valley as shown. The minimum merit location lies at the point (1.0, 1.0); however, if a starting value in the second quadrant is selected, convergence can sometimes be difficult to achieve. The behavior of the method of steepest descent will be demonstrated on this special merit function. A BASIC program that implements this solution follows.

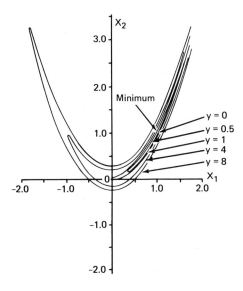

```
100   GOTO 9000
101 :
500   REM ***********************
510   REM * MERIT FUNCTION FOR  *
520   REM * THE ROSENBROCK MERIT*
530   REM * SURFACE.            *
540   REM ***********************
550 :
560   F = 100 * (X(2) - X(1) ^ 2) ^ 2 + (1
      - X(1)) ^ 2
570 :
580   REM  LOAD GRADIENTS
590   G(1) =  - 400 * X(1) * (X(2) - X(1)
      ^ 2) - 2 * (1 - X(1))
600   G(2) = 200 * (X(2) - X(1) ^ 2)
610   RETURN
620 :
2000  REM ********************
2010  REM * THIS SUBROUTINE   *
2020  REM * APPLIES THE       *
2030  REM * METHOD OF STEEPEST*
2040  REM * DESCENT TO        *
2050  REM * FIND THE UNCON-   *
2060  REM * STRAINED MINIMUM  *
2070  REM * OF A MERIT FUNC-  *
2080  REM * TION.             *
2090  REM *                   *
2100  REM * THE USER MUST     *
2110  REM * SUPPLY A STARTING *
2120  REM * VECTOR X(I), AN   *
2130  REM * ESTIMATE OF THE   *
2140  REM * OPTIMUM "EST" AND *
2150  REM * THE SIZE OF THE   *
2160  REM * NORMALIZED        *
2170  REM * GRADIENT VECTOR AT*
2180  REM * CONVERGENCE "EPS."*
2190  REM *                   *
2200  REM *                   *
2210  REM * PARAMETERS:       *
```

```
2220   REM *                      *
2230   REM *  NV  - THE NUMBER *
2240   REM *       OF DESIGN    *
2250   REM *       VARIABLES.   *
2260   REM *                      *
2270   REM *                      *
2280   REM * 500 - SUBROUTINE  *
2290   REM *       TO EVALUATE*
2300   REM *       THE MERIT    *
2310   REM *       VALUE "F"    *
2320   REM *       AND THE      *
2330   REM *       GRADIENT     *
2340   REM *       VECTOR G(I)*
2350   REM *       FOR A GIVEN*
2360   REM *       DESIGN VEC-*
2370   REM *       TOR "X(I)"  *
2380   REM *                      *
2390   REM * EPS -  TEST VALUE  *
2400   REM *       FOR THE      *
2410   REM *       SIZE OF THE*
2420   REM *       NORMALIZED  *
2430   REM *       GRADIENT AT*
2440   REM *       CONVERGENCE*
2450   REM *                      *
2460   REM *                      *
2470   REM * EST  - ESTIMATE OF*
2480   REM *       MINIMUM      *
2490   REM *       FUNCTION     *
2500   REM *       VALUE.       *
2510   REM *                      *
2520   REM *  FB  - ON RETURN   *
2530   REM *       THE BEST     *
2540   REM *       MERIT VALUE*
2550   REM *       FOUND.       *
2560   REM *                      *
2570   REM * XB(I)- THE DESIGN  *
2580   REM *       VECTOR COR-*
2590   REM *       RESPONDING   *
2600   REM *       TO FB.       *
2610   REM *                      *
2620   REM ********************
2630 :
2640   GOSUB 500
2650   REM   SARTUP
2660   REM   SET INITIAL DIRECTION
2670   REM   OF SEARCH
2680   FOR J = 1 TO NV
2690   P(J) =   - G(J)
2700   NEXT J
2710 :
2720 :
2730   REM   DO ONE-D SEARCH
2740   GOSUB 4000
2750 :
2760   REM   TEST FOR CONVERGENCE
2770   GG = 0!
2780   FOR J = 1 TO NV
2790   GG = GG + G(J) ^ 2
2800   NEXT J
2810 :
2820   REM    IF CONVERGED, FINISH
2830   IF GG < EPS GOTO 2890
2840 :
2850   REM   IF NOT CONVERGED THEN
2860   REM   START AGAIN
```

```
2870  GOTO 2640
2880  :
2890  REM   FINISH THE ANSWER
2900  FB = F
2910  FOR J = 1 TO NV
2920  XB(J) = X(J)
2930  NEXT J
2940  RETURN
2950  :
4000  REM   ********************
4010  REM   * THIS SUBROUTINE  *
4020  REM   * CONDUCTS A LINEAR*
4030  REM   * SEARCH TO FIND   *
4040  REM   * THE BEST VALUE   *
4050  REM   * IN A PARTICULAR  *
4060  REM   * DIRECTION        *
4070  REM   ********************
4080  :
4090  YB = F:VB = 0:PP = 0
4100  FOR J = 1 TO NV
4110  VB = VB + G(J) * P(J)
4120  PP = PP + P(J) * P(J)
4130  NEXT J
4140  IF VB >  = 0 GOTO 4640
4150  K = 2 * (EST - F) / VB
4160  :
4170  IF K < 0 GOTO 4220
4180  :
4190  IF K * K * PP > 1 GOTO 4220
4200  H = K
4210  GOTO 4230
4220  H = 1 /  SQR (PP)
4230  :
4240  REM   EXTRAPOLATION
4250  YA = YB:VA = VB
4260  FOR J = 1 TO NV
4270  X(J) = X(J) + H * P(J)
4280  NEXT J
4290  GOSUB 500
4300  YB = F:VB = 0
4310  FOR JJ = 1 TO NV
4320  VB = VB + G(JJ) * P(JJ)
4330  NEXT JJ
4340  IF VB >  = 0 GOTO 4410
4350  IF YB > YA GOTO 4410
4360  :
4370  H = H + K
4380  K = H
4390  GOTO 4250
4400  :
4410  T = 0
4420  :
4430  REM   INTERPOLATION
4440  Z = 3 * (YA - YB) / H + VA + VB
4450  W =  SQR (Z * Z - VA * VB)
4460  K = H * (VB + W - Z) / (VB - VA + 2
      * W)
4470  FOR J = 1 TO NV
4480  X(J) = X(J) + (T - K) * P(J)
4490  NEXT J
4500  GOSUB 500
4510  IF F > YA GOTO 4540
4520  IF F > YB GOTO 4540
4530  GOTO 4640
4540  VC = 0
```

```
4550   FOR JJ = 1 TO NV
4560   VC = VC + G(JJ) * P(JJ)
4570   NEXT JJ
4580   IF VC > 0 GOTO 4610
4590   YA=F:VA=VC:H=K:T=H
4600   GOTO 4620
4610   YB = F:VB = VC:H = H - K:T = 0
4620   GOTO 4440
4630  :
4640   RETURN
4650  :
9000   REM *********************
9010   REM * DRIVER PROGRAM    *
9020   REM *********************
9030  :
9040  NV = 2:EST = 0!
9050  EPS = .000001
9060   DIM X(NV),G(NV),P(NV),XB(NV)
9070  X(1) = -1.2:  X(2) = 1
9080  GOSUB 2000
9090  PRINT "---------------------------"
9100  PRINT "ANSWER IS";" F=";FB
9110  PRINT "X(1)=";X(1)
9120  PRINT "X(2)=";X(2)
9130  END
```

The subroutine starting at line 2000 in this program is based on
the method of steepest descent. The algorithm uses a special scheme
proposed by Davidon to perform the one-dimensional search for an
optimum along the gradient line. This special search scheme is done
in two phases. In the first phase extrapolation moves are made to
bracket the location of the one-dimensional optimum. Then an interpolation scheme based on a second-order polynomial approximation
to the function is used to locate the best possible value within the
brackets.

The output of this program is as follows:

```
---------------------------
ANSWER IS F= 9.536786E-07
X(1)= .9990247
X(2)= .9980454
```

This output required 128 seconds of run time on an IBM PC. During
this search, 325 different one-dimensional searches were conducted
and the merit subroutine was called 999 times. This performance is
an indication that the method is not well suited for this type of problem. If the user were to monitor the progress of the program through
a print statement in the merit function subroutine, the zigzag behavior characteristic of a ridge problem would be seen. Later in this
chapter we will see how an accelerated climbing technique can be
used to make a great improvement on the solution of this problem.

5-3 FLETCHER–REEVES METHOD

The Fletcher–Reeves method is an optimization algorithm for finding the unconstrained minimum of a multivariable nonlinear merit function of the form

$$\text{merit} = F(x_1, x_2, \ldots, x_n)$$

The method uses derivatives of the merit function with respect to the independent
variables. Unimodality is assumed, and therefore when using this method, the designer should try multiple starting points if multimodality is suspected. The logic
diagrams for this method is shown in Figure 5-2. The algorithm proceeds as follows. First, a feasible starting point in the design space is selected and the direction
of steepest descent is calculated in terms of the vector components

$$v_i^{(k)} = \frac{-\dfrac{\partial F^{(k)}}{\partial x_i}}{\left[\displaystyle\sum_{j=1}^{n}\left(\frac{\partial F}{\partial x_j}\right)^2\right]^{1/2}} \qquad i = 1, 2, \ldots, n$$

where $k = 1$ for the starting point. Next, a one-dimensional search is conducted
along the direction of steepest descent using the relationship

$$x_i(\text{new}) = x_i(\text{old}) + Sv_i \qquad i = 1, 2 \ldots, n$$

where S represents the distance moved along the gradient vector. When a minimum
is reached in this one-dimensional search, new search directions are computed.
These new directions are not exactly along the new gradient vector but are composed of a linear combination of the present gradient and the previous gradient. The
new direction vector components are found from

$$v_i^{(k+1)} = \frac{-\,\partial F^{(k+1)}/\partial x_i + B^{(k)}v_i^{(k)}}{\left\{\displaystyle\sum_{j=1}^{n}[-(\partial F^{(k+1)}/\partial x_j + B^{(k)}v_j^{(k)}]^2\right\}^{1/2}} \qquad i = 1, 2, \ldots, n$$

where

$$B^{(k)} = \frac{\displaystyle\sum_{i=1}^{n}[(\partial F^{(k+1)}/\partial x_i)]^2}{\displaystyle\sum_{i=1}^{n}[(\partial F^{(k)}/\partial x_i)]^2}$$

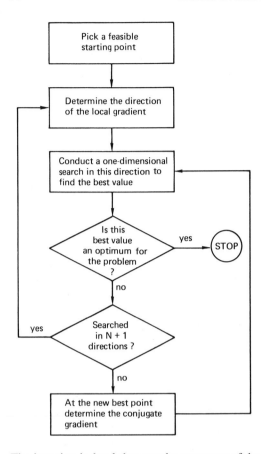

Figure 5-2   The Fletcher–Reeves algorithm.

The iteration index k denotes the sequence of the calculation in the iteration process. The new directions are said to be "conjugate" directions and are those directions corresponding to the current local quadratic approximation to the function. A one-dimensional search is then conducted long this new direction and when a minimum is found, a convergence check is made. If the check reveals that convergence is achieved, the procedure stops. If convergence is not achieved, a new set of conjugate directions is calculated, k is indexed by 1 and the process is repeated until convergence is achieved or until $n + 1$ directions have been used, at which point a new cycle is started using the steepest descent direction once again. The rationale for this algorithm is based on the prospect of exploiting the propensity that gradient techniques have for finding ridges in the design space. Since the second through $n + 1$ search directions are not along the gradient, they do not tend to "hang up" on the ridge but rather tend to exploit the notion that a ridge will usually point toward the extremum. It is true in general that methods which calculate a new search direction based on accumulated knowledge of the local behavior of the function are inherently more powerful than those in which the directions are assigned in advance. For this reason, the Fletcher–Reeves method is far superior to steepest ascent or descent methods. The disadvantage of the method is that since it is more complex than steepest gradient methods, it usually requires a correspondingly more complex subroutine for implementation.

Before proceeding to other gradient methods, let us illustrate the use of the Fletcher–Reeves method by an example application.

EXAMPLE 5-2

Suppose that it is desired to solve the problem of Example 5-1 using the method of Fletcher–Reeves. A BASIC program that implements this problem follows.

```
100  GOTO 9000
101  :
500  REM ************************
510  REM * MERIT FUNCTION FOR  *
520  REM * THE ROSENBROCK MERIT*
530  REM * SURFACE.            *
540  REM ************************
550  :
560  F = 100 * (X(2) - X(1) ^ 2) ^ 2 + (1
     - X(1)) ^ 2
570  :
580  REM  LOAD GRADIENTS
590  G(1) =  - 400 * X(1) * (X(2) - X(1)
     ^ 2) - 2 * (1 - X(1))
600  G(2) = 200 * (X(2) - X(1) ^ 2)
610  RETURN
620  :
2000  REM ********************
2010  REM * THIS SUBROUTINE   *
2020  REM * APPLIES THE       *
2030  REM * FLETCHER REEVES   *
2040  REM * ALGORITHM TO      *
2050  REM * FIND THE UNCON-   *
2060  REM * STRAINED MINIMUM  *
2070  REM * OF A MERIT FUNC-  *
2080  REM * TION.             *
2090  REM *                   *
2100  REM * THE USER MUST     *
2110  REM * SUPPLY A STARTING *
2120  REM * VECTOR X(I), AN   *
2130  REM * ESTIMATE OF THE   *
2140  REM * OPTIMUM "EST" AND *
2150  REM * THE SIZE OF THE   *
2160  REM * NORMALIZED        *
2170  REM * GRADIENT VECTOR AT*
2180  REM * CONVERGENCE "EPS."*
2190  REM *                   *
2200  REM *                   *
2210  REM * PARAMETERS:       *
2220  REM *                   *
2230  REM *  NV  - THE NUMBER *
2240  REM *        OF DESIGN  *
2250  REM *        VARIABLES. *
2260  REM *                   *
2270  REM *                   *
```

```
2280   REM * 500   - SUBROUTINE *
2290   REM *           TO EVALUATE*
2300   REM *           THE MERIT  *
2310   REM *           VALUE "F"  *
2320   REM *           AND THE    *
2330   REM *           GRADIENT   *
2340   REM *           VECTOR G(I)*
2350   REM *           FOR A GIVEN*
2360   REM *           DESIGN VEC-*
2370   REM *           TOR "X(I)" *
2380   REM *                      *
2390   REM * EPS   - TEST VALUE *
2400   REM *           FOR THE    *
2410   REM *           SIZE OF THE*
2420   REM *           NORMALIZED *
2430   REM *           GRADIENT AT*
2440   REM *           CONVERGENCE*
2450   REM *                      *
2460   REM *                      *
2470   REM * EST   - ESTIMATE OF*
2480   REM *           MINIMUM    *
2490   REM *           FUNCTION   *
2500   REM *           VALUE.     *
2510   REM *                      *
2520   REM *  FB   - ON RETURN   *
2530   REM *           THE BEST   *
2540   REM *           MERIT VALUE*
2550   REM *           FOUND.     *
2560   REM *                      *
2570   REM * XB(I)- THE DESIGN  *
2580   REM *           VECTOR COR-*
2590   REM *           RESPONDING *
2600   REM *           TO FB.     *
2610   REM *                      *
2620   REM ********************
2630 :
2640 :
2650   GOSUB 500
2660 N1 = NV + 1
2670   FOR I = 1 TO N1
2680 :
2690   REM   TEST FOR CONVERGENCE
2700 GG = 0
2710   FOR J = 1 TO NV
2720 GG = GG + G(J) * G(J)
2730   NEXT J
2740   IF GG < EPS GOTO 2980
2750 :
2760   REM   FIND SEARCH DIRECTION
2770   REM   START WITH STEEPEST
2780   REM   DESCENT THEN USE
2790   REM   CONJUGATE GRADIENT
2800   IF I > 1 GOTO 2860
2810   FOR J = 1 TO NV
2820 P(J) =  - G(J)
2830   NEXT J
2840   GOTO 2910
2850 :
2860 BETA = GG / OG
2870   FOR J = 1 TO NV
2880 P(J) =  - G(J) + BETA * P(J)
2890   NEXT J
2900 :
2910   REM   ONE-D SEARCH
```

```
2920  GOSUB 4000
2930  :
2940 OG = GG
2950  NEXT I
2960  GOTO 2670
2970  :
2980  :
2990  REM   FINISH THE ANSWER
3000 FB=F
3010 FOR J=1 TO NV
3020 XB(J) = X(J)
3030  NEXT J
3040  RETURN
3050  :
4000  REM   ********************
4010  REM   * THIS SUBROUTINE  *
4020  REM   * CONDUCTS A LINEAR*
4030  REM   * SEARCH TO FIND   *
4040  REM   * THE BEST VALUE   *
4050  REM   * IN A PARTICULAR  *
4060  REM   * DIRECTION        *
4070  REM   ********************
4080  :
4090 YB = F:VB = 0:PP = 0
4100  FOR J = 1 TO NV
4110 VB = VB + G(J) * P(J)
4120 PP = PP + P(J) * P(J)
4130  NEXT J
4140  IF VB >  = 0 GOTO 4640
4150 K = 2 * (EST - F) / VB
4160  :
4170  IF K < 0 GOTO 4220
4180  :
4190  IF K * K * PP > 1 GOTO 4220
4200 H = K
4210  GOTO 4230
4220 H = 1 /  SQR (PP)
4230  :
4240  REM   EXTRAPOLATION
4250 YA = YB:VA = VB
4260  FOR J = 1 TO NV
4270 X(J) = X(J) + H * P(J)
4280  NEXT J
4290  GOSUB 500
4300 YB = F:VB = 0
4310  FOR JJ = 1 TO NV
4320 VB = VB + G(JJ) * P(JJ)
4330  NEXT JJ
4340  IF VB >  = 0 GOTO 4410
4350  IF YB > YA GOTO 4410
4360  :
4370 H = H + K
4380 K = H
4390  GOTO 4250
4400  :
4410 T = 0
4420  :
4430  REM   INTERPOLATION
4440 Z = 3 * (YA - YB) / H + VA + VB
4450 W =  SQR (Z * Z - VA * VB)
4460 K = H * (VB + W - Z) / (VB - VA + 2
     * W)
4470  FOR J = 1 TO NV
4480 X(J) = X(J) + (T - K) * P(J)
```

```
4490  NEXT J
4500  GOSUB 500
4510   IF F > YA GOTO 4540
4520   IF F > YB GOTO 4540
4530   GOTO 4640
4540  VC = 0
4550   FOR JJ = 1 TO NV
4560  VC = VC + G(JJ) * P(JJ)
4570  NEXT JJ
4580  IF VC > 0 GOTO 4610
4590  YA = F:VA = VC:H = K:T = H
4600   GOTO 4620
4610  YB = F:VB = VC:H = H - K:T = 0
4620  GOTO 4440
4630  :
4640   RETURN
4650  :
9000   REM **********************
9010   REM * DRIVER PROGRAM      *
9020   REM **********************
9030  :
9040  NV = 2:EST = 0!
9050  EPS = .000001
9060   DIM X(NV),G(NV),P(NV),XB(NV)
9070  X(1) =   - 1.2:X(2) = 1
9080  GOSUB 2000
9090   PRINT "--------------------------"
9100   PRINT "ANSWER IS";" F=",FB
9110   PRINT "X(1)=";X(1)
9120   PRINT "X(2)=";X(2)
9130  END
```

The subroutine starting at line 2000 in this program is based on the method of Fletcher–Reeves. The algorithm uses the same one-dimensional search method used by the method of steepest descent. This routine begins at line 4000.

The output of this program is as follows:

```
--------------------------
ANSWER IS F=    1.381807E-09
X(1)= .9999635
X(2)= .9999278
```

This output required 11 seconds of run time on an IBM PC. During this search, 27 different one-dimensional searches were conducted and the merit subroutine was called only 57 times. This performance is an indication that this method is much better suited for this problem than is the method of steepest descent.

5-4 DAVIDON–FLETCHER–POWELL METHOD

The Davidon–Fletcher–Powell method is an optimization algorithm for finding the unconstrained minimum of a multivariable merit function of the form

$$\text{merit} = F(x_1, x_2, \ldots, x_n)$$

Derivatives of the merit function with respect to the independent variables are necessary. Since the algorithm is based on the assumption of unimodality, several alternative starting points are recommended if the merit surface is suspected to be multimodal. The logic diagram for this method is shown in Figure 5-3. The algorithm can be described as follows. First, a feasible point in design space is selected as a starting location. The direction of search is computed using the vector components

$$v_i^{(k)} = \frac{\sum\limits_{j=1}^{n} H_{i,j}(\partial F^{(k)}/\partial x_j)}{\left\{ \sum\limits_{l=1}^{n} \left[\sum\limits_{j=1}^{n} H_{l,j}(\partial F/\partial x_j) \right]^2 \right\}^{1/2}} \qquad i = 1, 2, \ldots, n$$

where k represents the iteration index and $H_{i,j}$ represents the elements of a symmetrical positive-definite $n \times n$ matrix. As the iteration process proceeds, this matrix becomes the inverse of the Hessian matrix (the matrix of second partial derivatives) of the merit function. Since this matrix is generally not known in advance, any symmetrical positive-definite matrix can be used to start the process. The usual choice is to use the identity matrix. When this particular selection is made, the initial direction of search is along the line of steepest descent. A one-dimensional search is conducted along this initial search direction using the following relation:

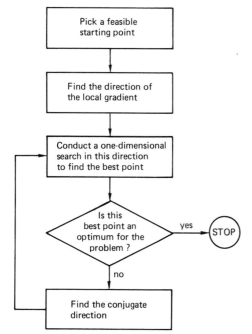

Figure 5-3   The Davidon–Fletcher–Powell algorithm.

$$x_i(\text{new}) = x_i(\text{old}) + Sv_i \qquad i = 1, 2, \ldots, n$$

where S is the step size in the direction of the search and v_i is the unit vector of steepest descent. A convergence check is made once the one-dimensional optimum is reached. If convergence is achieved, the search is terminated. If convergence is not achieved, a new search direction is chosen using the previous relationship and a new H matrix found as follows:

$$H^{(k+1)} = H^{(k)} + A^{(k)} - B^{(k)}$$

The elements of the $n \times n$ $A^{(k)}$ and $B^{(k)}$ matrices are computed by means of the following formulas (the superscript t denotes "transpose"):

$$A^{(k)} = \frac{\Delta x^{(k)} (\Delta x^{(k)})^t}{(\Delta x^{(k)})^t \, \Delta G^{(k)}}$$

$$B^{(k)} = \frac{H^{(k)} \, \Delta G^{(k)}(\Delta G^{(k)})^t H^{(k)}}{(\Delta G^{(k)})^t H^{(k)} \, \Delta G^{(k)}}$$

In these expressions $\Delta x^{(k)}$ and $\Delta G^{(k)}$ represent column vectors of differences in the x values and differences in the gradient values between locations. These column vectors are

$$\Delta x^{(k)} = x^{(k+1)} - x^{(k)} \qquad \text{(difference in location between iterations)}$$

$$\Delta G^{(k)} = \frac{\partial F^{(k+1)}}{\partial x} - \frac{\partial F^{(k)}}{\partial x} \qquad \text{(difference in gradients between iterations)}$$

Owing to the nature of the matrix operations, in the expressions for $A^{(k)}$ and $B^{(k)}$ the numerators are each $n \times n$ matrices and the denominators are scalars. Once the new search directions are known, a new one-dimensional search is performed, and the iterative process continues. By this algorithm the searches, after the first, are in directions of locally improving values of merit function but are rarely along the gradient. For this reason the algorithm is often called a *deflected gradient method*. Because of this special characteristic, the method is able to negotiate ridges in the design space. Most authorities regard this technique and the Fletcher–Reeves method as the most powerful of the gradient-based methods. Unlike the Fletcher–Reeves method, the Davidon–Fletcher–Powell method yields full information on the curvature of the merit function at any extremum found. Although this information is often useful, it is obtained at the price of providing storage space and manipulation time for the matrix H. For most microcomputer applications in optimization, the cost of time and storage space can be a significant factor in the selection of a particular algorithm. To illustrate the use of the Davidon–Fletcher–Powell method, let us now consider an example application.

EXAMPLE 5-3

Suppose that it is desired to solve the problem from Example 5-1 using
the method of Davidon–Fletcher–Powell. A BASIC program that implements this problem follows.

```
100   GOTO 9000
101 :
500   REM ************************
510   REM * MERIT FUNCTION FOR  *
520   REM * THE ROSENBROCK MERIT*
530   REM * SURFACE.            *
540   REM ************************
550 :
560 F = 100 * (X(2) - X(1) ^ 2) ^ 2 + (1
  - X(1)) ^ 2
570 :
580   REM  LOAD GRADIENTS
590 G(1) =  - 400 * X(1) * (X(2) - X(1)
^ 2) - 2 * (1 - X(1))
600 G(2) = 200 * (X(2) - X(1) ^ 2)
610   RETURN
620 :
2000   REM ********************
2010   REM * THIS SUBROUTINE   *
2020   REM * APPLIES THE       *
2030   REM * DAVIDON FLETCHER  *
2040   REM * POWELL ALGORITHM  *
2050   REM * ALGORITHM TO      *
2060   REM * FIND THE UNCON-   *
2070   REM * STRAINED MINIMUM  *
2080   REM * OF A MERIT FUNC-  *
2090   REM * TION.             *
2100   REM *                   *
2110   REM * THE USER MUST     *
2120   REM * SUPPLY A STARTING *
2130   REM * VECTOR X(I), AN   *
2140   REM * ESTIMATE OF THE   *
2150   REM * OPTIMUM "EST" AND *
2160   REM * THE SIZE OF THE   *
2170   REM * NORMALIZED        *
2180   REM * GRADIENT VECTOR AT*
2190   REM * CONVERGENCE "EPS."*
2200   REM *                   *
2210   REM *                   *
2220   REM * PARAMETERS:       *
2230   REM *                   *
2240   REM *  NV  - THE NUMBER *
2250   REM *         OF DESIGN *
2260   REM *         VARIABLES.*
2270   REM *                   *
2280   REM *                   *
2290   REM * 500  - SUBROUTINE *
2300   REM *         TO EVALUATE*
2310   REM *         THE MERIT *
2320   REM *         VALUE "F" *
2330   REM *         AND THE   *
2340   REM *         GRADIENT  *
2350   REM *         VECTOR G(I)*
2360   REM *         FOR A GIVEN*
```

```
2370  REM *           DESIGN VEC-*
2380  REM *           TOR "X(I)"  *
2390  REM *                       *
2400  REM * EPS -  TEST VALUE *
2410  REM *        FOR THE      *
2420  REM *        SIZE OF THE*
2430  REM *        NORMALIZED  *
2440  REM *        GRADIENT AT*
2450  REM *        CONVERGENCE*
2460  REM *                    *
2470  REM *                    *
2480  REM * EST  - ESTIMATE OF*
2490  REM *        MINIMUM     *
2500  REM *        FUNCTION    *
2510  REM *        VALUE.      *
2520  REM *                    *
2530  REM *  FB  - ON RETURN   *
2540  REM *        THE BEST    *
2550  REM *        MERIT VALUE*
2560  REM *        FOUND.      *
2570  REM *                    *
2580  REM * XB(I)- THE DESIGN  *
2590  REM *        VECTOR COR-*
2600  REM *        RESPONDING  *
2610  REM *        TO FB.      *
2620  REM *                    *
2630  REM * HH   - A 2-D ARRAY*
2640  REM *        ON RETURN   *
2650  REM *        IT CONTAINS*
2660  REM *        THE INVERSE*
2670  REM *        OF THE      *
2680  REM *        HESSIAN.    *
2690  REM *********************
2700 :
2710  GOSUB 500
2720  REM  SARTUP
2730  REM  SET HH TO IDENTITY
2740  REM  SET INITIAL DIRECTION
2750  REM  OF SEARCH
2760  REM  AND SET TEMP VALUES.
2770  FOR J = 1 TO NV
2780 P(J) =  - G(J)
2790 X1(J)=X(J):G1(J)=G(J)
2800  FOR JJ = 1 TO NV
2810 HH(J,JJ) = 0!
2820  NEXT JJ
2830 HH(J,J) = 1!
2840  NEXT J
2850 :
2860 :
2870 REM DO LINEAR SEARCH
2880 GOSUB 4000
2890 :
2900  REM  TEST FOR CONVERGENCE
2910 GG=0!
2920 FOR J=1 TO NV
2930 GG=GG+G(J)^2
2940 :
2950 NEXT J
2960  REM  IF CONVERGED, FINISH
2970 IF GG<EPS GOTO 3460
2980 :
2990  REM IF NOT CONVERGED THEN
3000 REM DETERMINE NEW
3010 REM SEARCH DIRECTION
```

```
3020 SY=0
3030 FOR J=1 TO NV
3040 YY(J)=G(J)-G1(J)
3050 SS(J)=X(J)-X1(J)
3060 SY = SY + YY(J) * SS(J)
3070  NEXT J
3080 :
3090  REM  IF SY=0 THEN RESTART
3100  REM  AT PRESENT BEST POINT
3110  REM  IN DIRECTION OF THE
3120  REM  STEEPEST DESCENT.
3130  IF SY = 0 THEN  GOTO 2710
3140 :
3150 BD = 0
3160  FOR J = 1 TO NV
3170 T = 0
3180 V1(J) = 0:V2(J) = 0
3190  FOR JJ = 1 TO NV
3200 T = T + YY(JJ) * HH(JJ,J)
3210 A(J,JJ) = SS(J) * SS(JJ) / SY
3220 V1(J) = V1(J) + HH(J,JJ) * YY(JJ)
3230 V2(J) = V2(J) + YY(JJ) * HH(JJ,J)
3240  NEXT JJ
3250 BD = BD + YY(J) * T
3260  NEXT J
3270 :
3280  FOR J = 1 TO NV
3290  FOR JJ = 1 TO NV
3300 B(J,JJ)=V1(J)*V2(JJ)/BD
3310 HH(J,JJ)=HH(J,JJ)+A(J,JJ)-B(J,JJ)
3320  NEXT JJ
3330  NEXT J
3340 :
3350  REM  NEW SEARCH DIRECTION
3360  REM  NEW TEMP VALUES
3370  FOR J = 1 TO NV
3380 X1(J) = X(J):G1(J) = G(J)
3390 P(J) = 0!
3400  FOR JJ = 1 TO NV
3410 P(J) = P(J) - HH(J,JJ) * G(JJ)
3420  NEXT JJ
3430  NEXT J
3440 GOTO 2870
3450 :
3460  REM  FINISH THE ANSWER
3470 FB = F
3480  FOR J = 1 TO NV
3490 XB(J) = X(J)
3500  NEXT J
3510  RETURN
3520 :
4000  REM  ********************
4010  REM  * THIS SUBROUTINE  *
4020  REM  * CONDUCTS A LINEAR*
4030  REM  * SEARCH TO FIND   *
4040  REM  * THE BEST VALUE   *
4050  REM  * IN A PARTICULAR  *
4060  REM  * DIRECTION        *
4070  REM  ********************
4080 :
4090 YB = F:VB = 0:PP = 0
4100  FOR J = 1 TO NV
4110 VB = VB + G(J) * P(J)
4120 PP = PP + P(J) * P(J)
4130  NEXT J
```

```
4140   IF VB >  = 0 GOTO 4640
4150   K = 2 * (EST - F) / VB
4160   :
4170   IF K < 0 GOTO 4220
4180   :
4190   IF K * K * PP > 1 GOTO 4220
4200   H = K
4210   GOTO 4230
4220   H = 1 /  SQR (PP)
4230   :
4240   REM   EXTRAPOLATION
4250   YA = YB:VA = VB
4260   FOR J = 1 TO NV
4270   X(J) = X(J) + H * P(J)
4280   NEXT J
4290   GOSUB 500
4300   YB = F:VB = 0
4310   FOR JJ = 1 TO NV
4320   VB=VB+G(JJ)*P(JJ)
4330   NEXT JJ
4340   IF VB >  = 0 GOTO 4410
4350   IF YB > YA GOTO 4410
4360   :
4370   H = H + K
4380   K = H
4390   GOTO 4250
4400   :
4410   T = 0
4420   :
4430   REM   INTERPOLATION
4440   Z = 3 * (YA - YB) / H + VA + VB
4450   W =  SQR (Z * Z - VA * VB)
4460   K = H * (VB + W - Z) / (VB - VA + 2
       * W)
4470   FOR J = 1 TO NV
4480   X(J) = X(J) + (T - K) * P(J)
4490   NEXT J
4500   GOSUB 500
4510   IF F > YA GOTO 4540
4520   IF F > YB GOTO 4540
4530   GOTO 4640
4540   VC = 0
4550   FOR JJ = 1 TO NV
4560   VC = VC + G(JJ) * P(JJ)
4570   NEXT JJ
4580   IF VC > 0 GOTO 4610
4590   YA = F:VA = VC:H = K:T = H
4600   GOTO 4620
4610   YB = F:VB = VC:H = H - K:T = 0
4620   GOTO 4440
4630   :
4640   RETURN
4650   :
9000   REM *********************
9010   REM * DRIVER PROGRAM     *
9020   REM *********************
9030   :
9040   NV = 2:EST = 0!
9050   EPS = .000001
9060   DIM X(NV),G(NV),P(NV),XB(NV)
9070   DIM HH(NV,NV),V1(NV),V2(NV)
9080   DIM A(NV,NV),B(NV,NV),YY(NV)
9090   X(1)=-1:X(2)=1
9100   DIM SS(NV),X1(NV),G1(NV)
9110 GOSUB 2000
```

```
9120 PRINT "-----------------------------"
9130 PRINT "ANSWER IS";" F=";FB
9140 PRINT "X(1)=";X(1)
9150 PRINT "X(2)=";X(2)
9160 END
```

The subroutine starting at line 2000 in this program is based on the method of Davidon–Fletcher–Powell. The algorithm uses the same one-dimensional search method used by the method of steepest descent. This routine begins at line 4000.

The output of this program is as follows:

```
-----------------------------
ANSWER IS F= 4.084768E-10
X(1)= .9999807
X(2)= .999962
```

This output required 37 seconds of run time on an IBM PC. During this search, 44 different one-dimensional searches were conducted and the merit subroutine was called only 108 times. This performance is an indication that the method is much better suited for this problem than is the method of steepest descent.

5-5 BROYDEN–FLETCHER–GOLDFARB–SHANNO METHOD

One of the more recent innovations in variable metric algorithms has been developed by several different researchers simultaneously. This method, called the *Broyden–Fletcher–Goldfarb–Shanno method,* is actually a modification of the Davidon–Fletcher–Powell method. The major difference in the method is that the formula used to update the H matrix is

$$H^{(k+1)} = H^{(k)} + A^{(k)} - B^{(k)}$$

where

$$A^{(k)} = \left[1 + \frac{(\Delta G^{(k)})^t H^{(k)} \, \Delta G^{(k)}}{(\Delta x^{(k)})^t \, \Delta G^{(k)}} \right] \frac{\Delta x^{(k)} (\Delta x^{(k)})^t}{(\Delta x^{(k)})^t \, \Delta G^{(k)}}$$

and

$$B^{(k)} = \frac{H^{(k)} \Delta G^{(k)} (\Delta x^{(k)})^t + \Delta x^{(k)} (\Delta G^{(k)})^t H^{(k)}}{(\Delta x^{(k)})^t \, \Delta G^{(k)}}$$

As before, in these expressions $\Delta x^{(k)}$ and $\Delta G^{(k)}$ represent column vectors of difference in the x_i values and differences in the gradient vector values between locations:

$$\Delta x^{(k)} = x^{(k+1)} - x^{(k)} \qquad \text{(difference in location between iterations)}$$

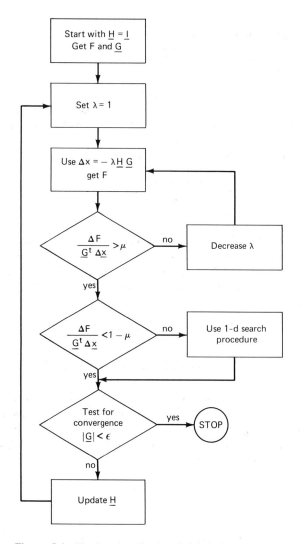

Figure 5-4 The Broyden–Fletcher–Goldfarb–Shanno Algorithm.

$$\Delta G^{(k)} = \frac{\partial F^{(k+1)}}{\partial x} - \frac{\partial F^{(k)}}{\partial x} \qquad \text{(difference in gradients between iterations)}$$

Although the expression for H is more complex than for the Davidon–Fletcher–Powell method, this formulation allows the use of $\Delta x^{(k)} = -H^k(\Delta G)^k$ as a means to move through the design space without the need for the Davidon linear search involving extrapolation and interpolation. This approach can lead to computational efficiency for complex problems because it requires fewer merit evaluations.

Fletcher reports that the abandonment of linear searches requires that something be done to force a sufficiently large improvement in F at each iteration to guarantee ultimate convergence. Such an approach would provide a scheme whereby any evaluation of the function and its gradient vector be made at each iteration. Fletcher reports that the change in F cannot become arbitrarily small if

$$\mu \leq \frac{\Delta F}{G^t \, \Delta x}$$

where $0 \leq \mu \leq 1$ is a preassigned small quantity. If corrections are determined by

$$\Delta x = -\lambda H^{(k)} G^{(k)}$$

then trying values of $\lambda = 1, W, W^2, W^3, \ldots$, where $0 < W < 1$, will eventually satisfy this test. Goldstein and Price (1967) have utilized $\mu = 0.0001$ and $W = 0.1$ in their work. These two researchers show that when converging to the minimum of a nonquadratic function, the choice $\lambda = 1$ will always be taken. To ensure convergence to a solution, it is also required that the step length λ not tend to zero. A sufficient condition for this is that

$$\frac{\Delta F}{G^t \, \Delta x} < 1 - \mu$$

Although values of λ in excess of 1 can be tried, this choice can lead to instability in the search process. A better choice might be to revert to the Davidon linear search scheme whenever values of λ in excess of 1.0 are called for by the condition above.

A diagram showing the Broyden–Fletcher–Goldfarb–Shanno algorithm is presented in Figure 5-4. For the sake of comparison we will now work an example application using this algorithm.

EXAMPLE 5-4

Suppose that it is desired to solve the problem from Example 5-1 using the Broyden–Fletcher–Goldfarb–Shanno method. A BASIC program that implements this problem follows.

```
100   GOTO 9000
101 :
500   REM ***********************
510   REM * MERIT FUNCTION FOR  *
520   REM * THE ROSENBROCK MERIT*
530   REM * SURFACE.            *
540   REM ***********************
550 :
560   F = 100 * (X(2) - X(1) ^ 2) ^ 2 + (1
      - X(1)) ^ 2
570 :
```

```
580  REM  LOAD GRADIENTS
590  G(1) =  - 400 * X(1) * (X(2) - X(1)
^ 2) - 2 * (1 - X(1))
600  G(2) = 200 * (X(2) - X(1) ^ 2)
610  RETURN
620  :
2000  REM *********************
2010  REM * THIS SUBROUTINE   *
2020  REM * APPLIES THE       *
2030  REM * BROYDEN, FLETCHER,*
2040  REM * GOLDFARB, SHANNO  *
2050  REM * ALGORITHM TO      *
2060  REM * FIND THE UNCON-   *
2070  REM * STRAINED MINIMUM  *
2080  REM * OF A MERIT FUNC-  *
2090  REM * TION.            *
2100  REM *                   *
2110  REM * THE USER MUST     *
2120  REM * SUPPLY A STARTING *
2130  REM * VECTOR X(I), AN   *
2140  REM * ESTIMATE OF THE   *
2150  REM * OPTIMUM "EST" AND *
2160  REM * THE SIZE OF THE   *
2170  REM * NORMALIZED        *
2180  REM * GRADIENT VECTOR AT*
2190  REM * CONVERGENCE "EPS."*
2200  REM *                   *
2210  REM *                   *
2220  REM * PARAMETERS:       *
2230  REM *                   *
2240  REM *  NV  - THE NUMBER *
2250  REM *         OF DESIGN *
2260  REM *         VARIABLES.*
2270  REM *                   *
2280  REM *                   *
2290  REM * 500  - SUBROUTINE *
2300  REM *         TO EVALUATE*
2310  REM *         THE MERIT *
2320  REM *         VALUE "F" *
2330  REM *         AND THE   *
2340  REM *         GRADIENT  *
2350  REM *         VECTOR G(I)*
2360  REM *         FOR A GIVEN*
2370  REM *         DESIGN VEC-*
2380  REM *         TOR "X(I)" *
2390  REM *                   *
2400  REM * EPS -   TEST VALUE *
2410  REM *         FOR THE   *
2420  REM *         SIZE OF THE*
2430  REM *         NORMALIZED *
2440  REM *         GRADIENT AT*
2450  REM *         CONVERGENCE*
2460  REM *                   *
2470  REM *                   *
2480  REM * EST  - ESTIMATE OF*
2490  REM *         MINIMUM   *
2500  REM *         FUNCTION  *
2510  REM *         VALUE.    *
2520  REM *                   *
2530  REM *  FB  - ON RETURN  *
2540  REM *         THE BEST  *
2550  REM *         MERIT VALUE*
2560  REM *         FOUND.    *
2570  REM *                   *
2580  REM * XB(I)- THE DESIGN *
```

```
2590  REM *          VECTOR COR-*
2600  REM *          RESPONDING *
2610  REM *          TO FB.      *
2620  REM *                      *
2630  REM * HH    - A 2-D ARRAY*
2640  REM *          ON RETURN   *
2650  REM *          IT CONTAINS*
2660  REM *          THE INVERSE*
2670  REM *          OF THE      *
2680  REM *          HESSIAN.    *
2690  REM *********************
2700 :
2710  GOSUB 500
2720  REM   SARTUP
2730  REM   SET HH TO IDENTITY
2740  REM   SET INITIAL DIRECTION
2750  REM   OF SEARCH
2760  FOR J = 1 TO NV
2770 P(J) =   - G(J)
2780  FOR JJ = 1 TO NV
2790 HH(J,JJ)=0
2800  NEXT JJ
2810 HH(J,J)=1
2820  NEXT J
2830 :
2840  REM   SET TEMP VALUES
2850  FOR J = 1 TO NV
2860 X1(J) = X(J):G1(J) = G(J)
2870 F1 = F
2880  NEXT J
2890 L=1
2900 :
2910  FOR J = 1 TO NV
2920 X(J) = X1(J) + L * P(J)
2930  NEXT J
2940 :
2950 GOSUB 500
2960 DF = F1 - F:DD = 0
2970 :
2980 REM ADJUST SIZE OF L TO INSURE
2990 REM REASONABLE CHANGES IN FUNCTION
VALUE
3000 IF ABS(DF/F1)<10 THEN GOTO 3020
3010 L=L*.1:GOTO 2910
3020 IF ABS(DF/F1)>.1 THEN GOTO 3050
3030 L=L*10:GOTO 2910
3040 :
3050  FOR J = 1 TO NV
3060 DD = DD + G(J) * L * P(J)
3070  NEXT J
3080 :
3090 IF ABS(DD)=0 THEN GOTO 3120
3100  IF ABS  (DF / DD) >  = .0001 THEN
 GOTO 3140
3110 :
3120 L = L * .1
3130  GOTO 2910
3140  IF ABS  (DF / DD) <  = .9999 THEN
 GOTO 3160
3150  GOSUB 4000
3160  REM   TEST FOR CONVERGENCE
3170 GG=0
3180  FOR J = 1 TO NV
3190 GG = GG + G(J) ^ 2
3200 :
```

```
3210  NEXT J
3220  REM   IF CONVERGED, FINISH
3230  IF GG < EPS GOTO 3720
3240  :
3250  :
3260  REM   IF NOT CONVERGED THEN
3270  REM   DETERMINE NEW
3280  REM   SEARCH DIRECTION
3290 SY = 0
3300 FOR J=1 TO NV
3310 YY(J)=G(J)-G1(J)
3320 SS(J)=X(J)-X1(J)
3330 SY=SY+YY(J)*SS(J)
3340 NEXT J
3350  :
3360 REM   IF SY=0 THEN RESTART
3370 REM   AT PRESENT BEST POINT
3380 REM   IN DIRECTION OF THE
3390 REM   STEEPEST DESCENT.
3400 IF SY=0 THEN GOTO 2710
3410  :
3420 BD=0
3430 FOR J=1 TO NV
3440 T=0
3450 V1(J)=0:V2(J)=0
3460 FOR JJ=1 TO NV
3470 T=T+YY(JJ)*HH(JJ,J)
3480 A(J,JJ)=SS(J)*SS(JJ)/SY
3490 V1(J)=V1(J)+HH(J,JJ)*YY(JJ)
3500 V2(J)=V2(J)+YY(JJ)*HH(JJ,J)
3510 NEXT JJ
3520 BD=BD+YY(J)*T
3530  NEXT J
3540  :
3550  FOR J = 1 TO NV
3560  FOR JJ = 1 TO NV
3570 B(J,JJ)=(V1(J)*SS(JJ)+V2(JJ)*SS(J))
/SY
3580 A(J,JJ)=A(J,JJ)*(1+BD/SY)
3590 HH(J,JJ) = HH(J,JJ) + A(J,JJ) - B(J
,JJ)
3600  NEXT JJ
3610  NEXT J
3620  :
3630  REM   NEW SEARCH DIRECTION
3640  REM   NEW TEMP VALUES
3650  FOR J = 1 TO NV
3660  P(J)=0
3670  FOR JJ = 1 TO NV
3680  P(J) = P(J) - HH(J,JJ) * G(JJ)
3690  NEXT JJ
3700  NEXT J
3710  GOTO 2850
3720  :
3730  REM   FINISH THE ANSWER
3740 FB = F
3750  FOR J = 1 TO NV
3760 XB(J) = X(J)
3770  NEXT J
3780  RETURN
3790  :
4000  REM   ********************
4010  REM   * THIS SUBROUTINE  *
4020  REM   * CONDUCTS A LINEAR*
4030  REM   * SEARCH TO FIND   *
```

```
4040  REM  * THE BEST VALUE   *
4050  REM  * IN A PARTICULAR  *
4060  REM  * DIRECTION        *
4070  REM  *********************
4080 :
4090 YB = F:VB = 0:PP = 0
4100  FOR J = 1 TO NV
4110 VB = VB + G(J) * P(J)
4120 PP = PP + P(J) * P(J)
4130 NEXT J
4140  IF VB >  = 0 GOTO 4640
4150 K = 2 * (EST - F) / VB
4160 :
4170  IF K < 0 GOTO 4220
4180 :
4190  IF K * K * PP > 1 GOTO 4220
4200 H = K
4210  GOTO 4230
4220 H = 1 / SQR (PP)
4230 :
4240  REM  EXTRAPOLATION
4250 YA = YB:VA = VB
4260  FOR J = 1 TO NV
4270 X(J) = X(J) + H * P(J)
4280  NEXT J
4290  GOSUB 500
4300 YB = F:VB = 0
4310  FOR JJ = 1 TO NV
4320 VB=VB+G(JJ)*P(JJ)
4330 NEXT JJ
4340  IF VB >  = 0 GOTO 4410
4350  IF YB > YA GOTO 4410
4360 :
4370 H = H + K
4380 K = H
4390  GOTO 4250
4400 :
4410 T = 0
4420 :
4430  REM  INTERPOLATION
4440 Z = 3 * (YA - YB) / H + VA + VB
4450 W = SQR (Z * Z - VA * VB)
4460 K = H * (VB + W - Z) / (VB - VA + 2
     * W)
4470  FOR J = 1 TO NV
4480 X(J) = X(J) + (T - K) * P(J)
4490  NEXT J
4500  GOSUB 500
4510  IF F > YA GOTO 4540
4520  IF F > YB GOTO 4540
4530  GOTO 4640
4540 VC = 0
4550  FOR JJ = 1 TO NV
4560 VC = VC + G(JJ) * P(JJ)
4570  NEXT JJ
4580  IF VC > 0 GOTO 4610
4590 YA = F:VA = VC:H = K:T = H
4600  GOTO 4620
4610 YB = F:VB = VC:H = H - K:T = 0
4620  GOTO 4440
4630 :
4640  RETURN
4650 :
9000  REM  *********************
9010  REM  * DRIVER PROGRAM    *
```

```
9020   REM ***********************
9030   :
9040   NV=2:EST=0
9050   EPS = .000001
9060   DIM X(NV),G(NV),P(NV),XB(NV)
9065   DIM HH(NV,NV),V1(NV),V2(NV)
9066   DIM A(NV,NV),B(NV,NV),YY(NV)
9070   X(1)=-1.2:X(2)=1
9077   DIM SS(NV), G1(NV),X1(NV)
9080   GOSUB 2000
9090   PRINT "--------------------------"
9100   PRINT "ANSWER IS";" F=",FB
9110   PRINT "X(1)=";X(1)
9120   PRINT "X(2)=";X(2)
9130   END
```

The subroutine starting at line 2000 in this program is based on the Broyden–Fletcher–Goldfarb–Shanno method. The algorithm attempts to circumvent the need for a one-dimensional search whenever possible. If the use cannot be avoided, the Davidon one-dimensional search method is used. This routine begins at line 4000 and is the same routine that was used for Examples 5-1 to 5-3. The output of this program is as follows:

```
--------------------------
ANSWER IS F=    1.604406E-11
X(1)= .9999968
X(2)= .9999933
```

This output required 13 seconds of run time on an IBM PC. During this search only nine different one-dimensional searches were conducted. In a number of cases, the one-dimensional search was not needed and, in all, the merit subroutine was called only 40 times. This performance is an indication that the method is much better suited to this problem than is the method of steepest descent. The method requires fewer iterations than both the Fletcher–Reeves method and the Davidon–Fletcher–Powell method. A summary of the comparison of the four gradient-based methods is presented below. This table shows that the Broyden–Fletcher–Goldfarb–Shanno method would be best to use if the amount of effort required to perform the merit evaluations is significant. The table also shows the trade-off between the number of iterations required and the added complexity of methods that utilize the variable metric method.

Method	Seconds of run time for EPS = 0.000001	Number of merit evaluations used	Number of $1-d$ searches used	Bytes of space with remarks
Steepest descent	128	999	325	4965
Fletcher–Reeves	11	57	27	5152
Davidon–Fletcher–Powell	37	108	44	6455
Broyden–Fletcher–Goldfarb–Shanno	13	40	9	7061

5-6 CONSIDERATIONS IN THE SELECTION OF A GRADIENT METHOD

When performing multidimensional optimization on the microcomputer using gradient methods, several points are important to keep in mind. These points are classified in terms of method selection, problem setup, and program technique. Each of these critical points is now discussed.

Method Selection

In selecting a gradient method for optimization it is well to keep in mind that gradient methods tend to have trouble at boundaries and ridges. If the user suspects this type of topology for the merit surface, it is wise to avoid gradient methods altogether. If the topology is uncertain, the methods of conjugate gradient and the variable metric algorithms can sometimes be successful if the boundaries are not severe. Boundary failure of gradient methods can be recognized as a zigzag movement that makes little progress toward the actual optimum. When the merit function for a given problem is relatively simple, the method of Fletcher–Reeves is often the most efficient method to use. When the merit function is relatively complex and requires considerable computational effort to determine, the method of Broydon–Fletcher–Goldfarb–Shanno will usually reach a solution with the minimum of computer time. It should be kept in mind that the amount of storage space required by the variable metric methods goes up dramatically as the number of design variables goes up. This is because of the extra space required for the Hessian matrix, the A matrix, the B matrix, and intermediate vectors used to find the Hessian matrix. Thus if computer storage space is at a premium, the user may choose to use the Fletcher–Reeves method even though the variable metric methods may be a bit faster to converge. On the other hand, if information about the nature of the type of extremum found is required, only the variable metric methods provide this information as direct output. Because many engineering problems have solutions along boundaries where the local gradient is not zero, indirect optimization methods are seldom useful. When setting up a problem that is suspected to have multimodal behavior, it is sound practice to try a number of starting points within the design space to see if they all converge to the same optimum location.

Problem Setup

Many types of engineering problems have no closed form for the merit function, or the function may be so complex that differentiation is difficult. In such cases the derivative may be difficult or impossible to extract in exact form. For these cases the secant approximation to a derivative can be used. This method utilizes an approximation to the derivative found by dividing a change in merit value by a corresponding small change in a single independent variable. Although such a procedure can be helpful, it is well to keep in mind that it requires at least n extra merit evaluations to approximate each set of partial derivatives.

Program Technique

When preparing a program for gradient-based optimization it is well to include some provision in the program to monitor the progress of the optimization process. In this way the user can detect a boundary or ridge impasse and restart the method from some other feasible starting point. This procedure can save the designer considerable wasted effort for problems that have no potential to converge. Since the user of a personal computer has considerably more control over the execution of a program than the user of a mainframe computer or a minicomputer, careful preparation of an optimization program can provide considerable efficiency in the problem-solving process. Thus even a gradient-based program that fails to converge can often provide useful information to the designer if the program is constructed to show the intermediate steps of the optimization process.

PROBLEMS

5-1 An open-top box has 40 equal-sized compartments constructed with partitions in the 5 by 8 array shown. If the volume of each compartment must be 1000 cm³, find the dimensions a, b, and c so that a minimum of material is used for the design.

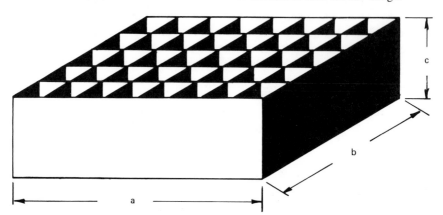

5-2 Rework Problem 5-1 assuming that the box has a lid.

5-3 The table in Example 5-4 compares the speed of convergence for four different gradient methods. How would the run time for each of these be different if the value of ϵ were changed from 0.000001 to 0.001?

5-4 Unlike the Rosenbrock merit function used for the examples in this chapter, the merit function

$$F = 2X_1^2 + X_2^2 + 3X_3^2$$

does not have a curved valley. Experiment with this merit function and the starting point $(-1, 1, -1)$ to see if the method of steepest descent compares more favorably with the method of Fletcher–Reeves.

5-5 Fletcher and Powell have studied an unusual merit surface with a helical valley in three dimensions:

$$F(x_1, x_2, x_3) = 100[(x_3 - 10\theta)^2 + (r - 1)^2] + x_3$$

where

$$x_1 = r \cos 2\pi\theta$$
$$x_2 = r \sin 2\pi\theta$$

if the value of θ is restricted to

$$-\tfrac{1}{4} \le \theta \le \tfrac{3}{4}$$

find a minimum from the starting point $(-1, 0, 0)$ using several different methods.

5-6 Experiment with the random search algorithm from Chapter 4 to see if you can find a better starting point for Problem 5-5 than $(-1, 0, 0)$.

5-7 Solve Example 4-1 using a gradient method.

5-8 An interesting variation on the Rosenbrock merit functions is

$$F = 100(x_2 - x_1^3)^2 + (1 - x_1)^2$$

Start with $(-1.2, 1.0)$ and find the minimum of this function using one of the methods presented in this chapter.

5-9 The merit function of Beale,

$$F = \sum_{i=1}^{3} [c_i - x_1(1 - x_2^i)]^2$$

with $c_1 = 1.4$, $c_2 = 2.25$, and $c_3 = 2.625$, has a narrow curving valley and a minimum at $(3, 0.5)$. Try several of the methods in this chapter to see which works best on this problem.

5-10 A 8.9-cm-OD 7.8-cm-ID steel ($k = 43.27$ W/m-°C) pipe transports a fluid at 148°C. The pipe is insulated with two layers of insulation, as shown. The inner layer of insulation has a thermal conductivity of $k = 0.20$ W/m-°C and the outer layer has a thermal conductivity of $k = 0.5$ W/m-°C. The coefficient of convection on the inside of the

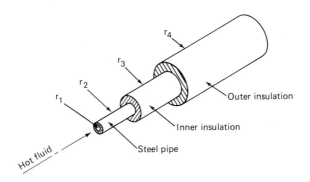

pipe is 230 W/m^2-°C and on the outside of the outer insulation it is 23 W/m^2-°C. Design the insulation layers for minimum cost if the maximum outer surface temperature allowed is 38°C and the ambient air temperature is 27°C. The maximum allowable outer radius is 12 cm. The cost of the inner insulation material is $35/m^3$, and the cost of the outer material is $100/m^3$. (You may neglect the thermal contact loss between layers of material.)

REFERENCES

1. Fletcher, R., and Powell, M. J. D., "A Rapidly Convergent Descent Method for Minimization," *Computer Journal,* Vol. 6, 1963, pp. 163–168.
2. Fletcher, R., and Reeves, C. M., "Function Minimization by Conjugate Gradients," *Computer Journal,* Vol. 7, 1964, pp. 149–154.
3. Fletcher, R., ed., *Optimization,* Academic Press, Inc., New York, 1969.
4. Shoup, T. E., *A Practical Guide to Computer Methods for Engineers,* Prentice-Hall, Inc., Englewood Cliffs, N.J., 1979.
5. Shoup, T. E., *Numerical Methods for the Personal Computer,* Prentice-Hall, Inc., Englewood Cliffs, N.J., 1983.
6. Wilde, D. J., and Beightler, C. S., *Foundations of Optimization,* Prentice-Hall, Inc., Englewood Cliffs, N.J., 1967.
7. Mayne, R. W., and Ragsdell, K. M., *Progress in Engineering Optimization,* American Society of Mechanical Engineers, New York, 1981.
8. Goldstein, A. A., and Price, J.F., "An Effective Algorithm for Minimizations," *Numerische Mathematik,* Vol. 10, 1967, 184–189.
9. Broyden, G. G., "The Convergence of a Class of Double Rank Minimization Algorithms, the New Algorithm," *Journal of the Institute of Mathematics and Its Applications,* Vol. 6, 1970, pp. 222–231.
10. Fletcher, R., "A New Approval to Variable Metric Algorithms," *Computer Journal,* Vol. 13, 1970, pp. 317–322.
11. Goldfarb, D., "A Family of Variable Metric Methods Derived by Variational Means," *Mathematics of Computation,* Vol. 24, 1970, pp. 23–26.
12. Shanno, D. F., "Conjugate Gradient Methods with Inexact Searches," *Mathematics of Operations Research,* Vol. 3, 1978, pp. 244–256.

6

PATTERN SEARCH METHODS FOR MULTIDIMENSIONAL OPTIMIZATION

With the use of proper peripheral equipment, the personal computer can become a versatile tool for communicating ideas in graphical form. (Courtesy of Calcomp.)

A large number of multidimensional optimization problems in engineering have merit functions that are mathematically complex or are based on tabular data. For these optimization problems, finding the necessary partial derivatives required to assemble the gradient is often impossible. In these situations it is necessary to use a search algorithm that does not depend on calculation of the gradient. One type of technique that satisfies this requirement is the *pattern search method*. Pattern search methods are known for their simplicity and are popular in the optimization field because their performance is often better than methods that rely on gradient calculation or gradient approximation. Pattern search algorithms have been demonstrated to be far superior to gradient methods when used on merit surfaces that have sharply defined ridges as a result of the imposed constraints.

In this chapter we discuss the characteristics of several of the commonly used pattern search methods. Example software and applications for implementation on the personal computer are also presented to illustrate the relative ease of use of these powerful techniques.

6-1 HOOKE AND JEEVES PATTERN SEARCH METHOD

The pattern search technique of Hooke and Jeeves is an easily programmed, climbing technique that does not require the use of derivatives. The algorithm has ridge-following properties and is based on the premise that any set of design moves that have been successful during early experiments will be worth trying again. The method is based on the assumption of unimodality and is used to find the minimum of a multivariable, unconstrained function of the form

$$\text{merit} = F(x_1, x_2, \ldots, x_n)$$

The logic diagram for this method is shown in Figure 6-1. The algorithm proceeds as follows. First, a base point in the feasible design space is chosen together with exploration step sizes. Next, an exploration is performed a given increment along each of the independent-variable directions following the logic shown in Figure 6-2. Whenever a functional improvement is obtained, a new temporary base point is established. Once this exploration is complete, a new base point is established and a "pattern move" takes place. This pattern move consists of an extrapolation along a line between the new base point and the previous base point. The distance moved beyond the best base point is somewhat larger than the distance between the two base points. Mathematically, this extrapolation is

$$x_{i,0}^{(k+1)} = x_i^{(k+1)} + a(x_i^{(k+1)} - x_i^{(k)})$$

where $x_{i,0}^{(k+1)}$ becomes a new temporary base point or "head." In this expression i is the variable index, k is the stage index, and a is an acceleration factor that is greater than or equal to 1.0. Once the new temporary base point has been found, an exploration about this point is instituted to see if a better base point can be found. This exploration also uses the logic of Figure 6-2. If the temporary head or any of its

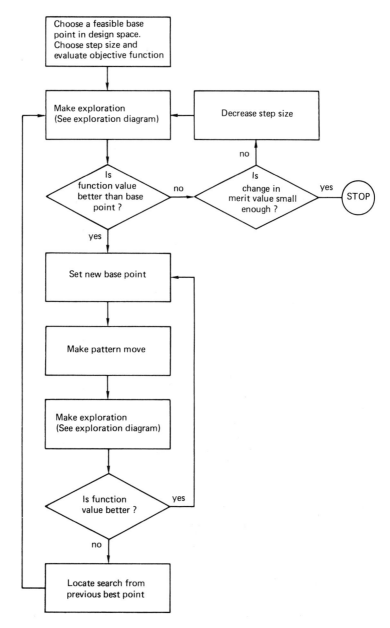

Figure 6-1 The Hooke and Jeeves pattern search algorithm.

neighboring points is a better base, the pattern process repeats using this improved base. Because of the nature of the acceleration factor, each successive pattern extrapolation becomes bolder and bolder until the process oversteps the peak or a ridge. At this point the previous "best base" is recalled, the local exploration step

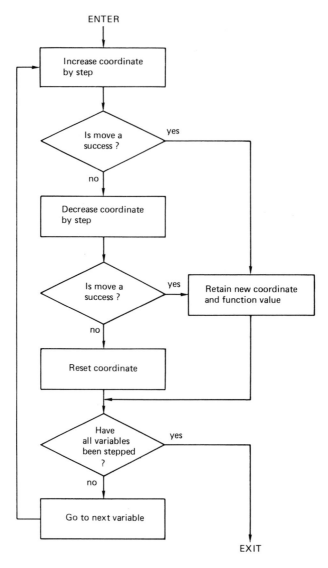

Figure 6-2 The exploration method used in the Hooke and Jeeves algorithm.

size is decreased, and the pattern-building process begins again. Once the step size is decreased below a predetermined value and still no substantial change in the merit value can be achieved, the procedure terminates. Although this method lacks mathematical elegance, it is a highly efficient optimization algorithm. The relative simplicity of the method has made it one of the most popular for computer implementation and this simplicity makes it an ideal candidate for use on the small computer.

It should be pointed out that the pattern search method is not entirely foolproof. Under certain circumstances it can terminate at a false optimum. An example of this situation is shown in Figure 6-3 for the case of two design variables. In this case the method has found a "best" base point that is on a constraint boundary. Any moves in the plus x_2 direction or the $-x_1$ direction will be penalized so that they are not improved values used to move toward the minimum. Explorations taken in the opposite directions also do not yield improvement no matter how small the exploration step size becomes. The method is therefore stalled on the boundary. This type of behavior can be recognized by trying a different starting point to see if the method terminates at another location along the boundary. When this type of situation is encountered, the particular problem calls for a different type of penalty function that will not allow the merit value to come to rest on the boundary line. Such methods are based on solving a sequence of penalized merit functions that have an infinite merit value along the boundary. As the process proceeds, the influence of the penalty part of the merit function is reduced in a sequential fashion. In the end the problem is solved without any penalty function to obtain the final solution. The details of this method were discussed in Chapter 2. Another approach to overcome this pitfall is to switch to another method.

Some researchers recommend that the test for convergence in the Hooke and Jeeves method be replaced by a random search in the vicinity of the best point. If this random search produces a more favorable merit value, the search is not termi-

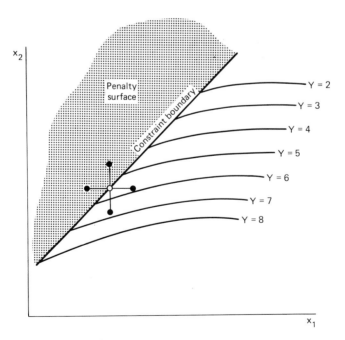

Figure 6-3   Failure of the Hooke & Jeeves pattern search in the neighborhood of a constraint boundary.

nated but is restarted from this new base point. This approach will be left as an exercise for the reader to implement.

It should be mentioned that the stalled behavior described above can occur at points away from the penalty boundaries if the merit surface has a sharp ridge with merit contours that are slanted 45 degrees relative to the directions of the design variable axes. This particular pitfall can sometimes be overcome if the directions of the exploratory searches are allowed to change orientation so that they become normal to the merit contours. Although this approach can provide improvement, it may still fail to remedy the problem entirely. Nevertheless, this type of search algorithm is worthy of consideration. One of the commonly used methods of this type is the Rosenbrock pattern search method. This method is described in Section 6-2. Before proceeding with this method, however, let us consider an example application of the Hooke and Jeeves method.

EXAMPLE 6-1

The test problem by Colville is a four-dimensional version of Rosenbrock's test function solved in Chapter 5. The merit function for this problem is

$$F = 100(x_2 - x_1^2)^2 + (1 - x_1)^2$$
$$+ 90(x_4 - x_3^2)^2 + (1 - x_3)^2$$
$$+ 10.1[(x_2 - 1)^2 + (x_4 - 1)^2]$$
$$+ 19.8(x_2 - 1)(x_4 - 1)$$

The constraints are

$$-10 \le x_i \le 10 \qquad i = 1, 2, 3, 4$$

A feasible starting point is

$$x_1 = -3$$
$$x_2 = -1$$
$$x_3 = -3$$
$$x_4 = -1$$

with $F = 19,192.0$. The problem has a global optimum at

$$x_i = 1 \qquad i = 1, 2\ 3, 4$$

This problem also has a local optimum at $(-1, 1, -1, 1)$. For this location the merit value is $F = 7.876$.

Let us prepare and run a computer program that will find the global optimum using the method of Hooke and Jeeves. Since the problem is constrained, we will convert it into an unconstrained problem by means of a penalty surface that sets $F = 20,000$ whenever any one of the eight constraint inequalities is violated. This choice is based on the assumption that the value is worse than the starting point and thus must be worse than the optimum value. Since the method does not depend in any way on the use of a derivative, a penalty surface that is absolutely flat will probably suffice. A BASIC program that implements this solution follows.

```
100   GOTO 9000
110 :
500   REM ***********************
510   REM * MERIT FUNCTION FOR  *
520   REM * COLVILLE'S MERIT     *
530   REM * SURFACE              *
540   REM ***********************
550 :
560 F=100*(X(2)-X(1)^2)^2+(1-X(1))^2+90*
(X(4)-X(3)^2)          ^2+(1-X(3))^2+10.1
*((X(2)-1)^2+(X(4)-1)^2)+19.8*(X(2)-1)*(
X(4)-1)
570 :
580   REM   CHECK CONSTRAINTS
590   FOR IX = 1 TO NV
600   IF  ABS (X(IX)) > 10 GOTO 650
610   NEXT IX
620   RETURN
630 :
640   REM   PENALIZE
650 F = 20000!
660   RETURN
670 :
2000   REM *********************
2010   REM * THIS SUBROUTINE   *
2020   REM * APPLIES THE HOOKE *
2030   REM * & JEEVES ALGORITHM*
2040   REM * OF EXPLORATION AND*
2050   REM * PATTERN MOVES TO  *
2060   REM * FIND THE UNCON-   *
2070   REM * STRAINED MINIMUM  *
2080   REM * OF A MERIT FUNC-  *
2090   REM * TION.  THE USER   *
2100   REM * MUST SUPPLY A     *
2110   REM * STARTING VECTOR   *
2120   REM * X(I).             *
2130   REM *                   *
2140   REM * PARAMETERS:       *
2150   REM *                   *
2160   REM * NV  - THE NUMBER  *
2170   REM *       OF DESIGN   *
2180   REM *       VARIABLES.  *
2190   REM *                   *
2200   REM * R   - INITIAL     *
2210   REM *       EXPLORATION *
2220   REM *       STEP SIZE   *
2230   REM *                   *
```

```
2240  REM * A     - ACCELERA-  *
2250  REM *         TION FACTOR*
2260  REM *         SET >=1.0   *
2270  REM *                     *
2280  REM * RF    - FINAL       *
2290  REM *         EXPLORATION*
2300  REM *         STEP SIZE.  *
2310  REM *                     *
2320  REM * 500   - SUBROUTINE  *
2330  REM *         TO EVALUATE*
2340  REM *         THE MERIT    *
2350  REM *         VALUE "F"    *
2360  REM *         USING A      *
2370  REM *         DESIGN VEC-*
2380  REM *         TOR "X(I)"  *
2390  REM *                     *
2400  REM * FB    - ON RETURN   *
2410  REM *         THE BEST     *
2420  REM *         MERIT VALUE*
2430  REM *         FOUND.       *
2440  REM *                     *
2450  REM * XB(I)- THE DESIGN   *
2460  REM *         VECTOR COR-*
2470  REM *         RESPONDING  *
2480  REM *         TO FB.       *
2490  REM *                     *
2500  REM *********************
2510  :
2520  GOSUB 500
2530  FB=F
2540  R=R
2550  FOR N=1 TO NV
2560  X(N)=XB(N)
2570  NEXT N
2580  :
2590  REM EXPLORE
2600  GOSUB 2960
2610  :
2620  REM IS IT BETTER?
2630  REM IF SO GO TO PATTERN
2640  IF FE<FB THEN GOTO 2720
2650  :
2660  REM IF NOT DECREASE
2670  REM STEP SIZE DOWN TO RF
2680  IF R<RF THEN RETURN
2690  R=R/2!
2700  GOTO 2550
2710  :
2720  REM MAKE PATTERN MOVE
2730  FOR N=1 TO NV
2740  X(N)=XE(N)+A*(XE(N)-XB(N))
2750  NEXT N
2760  :
2770  REM REPLACE XB WITH XE
2780  FOR N=1 TO NV
2790  XB(N)=XE(N)
2800  NEXT N
2810  FB=FE
2820  :
2830  REM EXPLORE FROM HERE
2840  GOSUB 2960
2850  :
2860  REM IF ITS BETTER
2870  REM REPEAT PATTERN
2880  :
```

```
2890 IF FE<FB GOTO 2720
2900 :
2910 REM IF NOT GO BACK TO
2920 REM BEST BASE POINT
2930 REM AND EXPLORE
2940 GOTO 2540
2950 :
2960 REM ********************
2970 REM * THIS SUBROUTINE   *
2980 REM * PERFORMS THE      *
2990 REM * EXPLORATION STEP  *
3000 REM ********************
3010 :
3020 GOSUB 500:FF=F
3030 FE=F
3040 FOR N=1 TO NV
3050 XE(N)=X(N)
3060 X(N)=X(N)+R
3070 GOSUB 500
3080 IF F<FF GOTO 3140
3090 X(N)=X(N)-2*R
3100 GOSUB 500
3110 IF F<FF GOTO 3140
3120 X(N)=X(N)+R
3130 GOTO 3160
3140 FE=F:FF=FE
3150 XE(N)=X(N)
3160 NEXT N
3170 RETURN
3180 :
9000  REM ********************
9010  REM * DRIVER PROGRAM    *
9020  REM * PERFORMS AN UNCON-*
9030  REM * STRAINED OPTIMIZA-*
9040  REM * TION OF A MERIT   *
9050  REM * FUNCTION.         *
9060  REM ********************
9070 :
9080 RF=.0001:A=1.1
9090 NV=4:R=3
9100 DIM XB(NV),X(NV),XE(NV)
9110 :
9120 REM FEASIBLE START POINT
9130 X(1)=-3:X(2)=-1:X(3)=-3:X(4)=-1
9140 FOR IX=1 TO NV
9150 XB(IX)=X(IX)
9160 NEXT IX
9170 :
9180 REM CALL HOOKE & JEEVES
9190 GOSUB 2000
9200 :
9210 PRINT "FINAL RESULT IS":PRINT"F=";F
B:PRINT"X(1)=";XB(1):PRINT"X(2)=";XB(2)
9220 PRINT "X(3)=";XB(3)
9230 PRINT "X(4)=";XB(4)
9240 END
```

To run this program the user must set the values for the the acceleration factor A in the pattern search as well as the starting step size R for exploration and the end step size for exploration RF. The choices for A = 1.1, R = 3.0, and RF = 0.0001 are typical for the operation of the Hooke and Jeeves method. Small changes from these values will result in nearly the same performance for the method. If the starting step size R is changed by a significant amount it is possible to stumble on a track that will lead to the local optimum at $(-1, 1, -1, 1)$. The output of this program is as follows:

```
FINAL RESULT IS
F= 1.025326E-06
X(1)= 1.000028
X(2)= 1.000042
X(3)= .9998329
X(4)= .9996508
```

This output required 62 seconds to complete on an IBM PC. In performing this optimization, the program has made 535 merit evaluations. As it nears the optimum, this algorithm tends to slow down after having made rapid progress. Because of the significant level of time and effort required to achieve this optimum, the user would be well advised to monitor the progress of the search by including a print statement in the merit routine. In this way the user could terminate the program if the optimum that is being approached is not the global optimum desired. Space limitations do not allow printing the 535 iterations that this program uses to achieve a solution, but the interested user would be well advised to experiment with the program to see how the method progresses on a typical problem.

The added complexity of this problem due to the increased dimensionality of the design space makes this problem significantly more difficult to solve.

6-2 ROSENBROCK PATTERN SEARCH METHOD

The pattern search of Rosenbrock is a ridge-following method that has been shown to be effective often when other methods fail. This method is frequently called the *method of rotating coordinates* because of the way it performs the local explorations. Rather than perturbing each of the original variables independently, this method rotates the coordinate system so that one axis points along a ridge. The location of this ridge is determined by a previous trial. The remaining axes are arranged orthogonal to this first axis. The method is based on the assumption of unimodality and is used to find the minimum of a multivariable, unconstrained function of the form

$$\text{merit} = F(x_1, x_2, \ldots, x_n)$$

The logic diagram for this method is shown in Figure 6-4. The algorithm proceeds as follows. First a starting point and initial step sizes $(S_i, i = 1, 2, \ldots, n)$ are picked, and the merit function is evaluated. Then, in turn, each variable x_i is

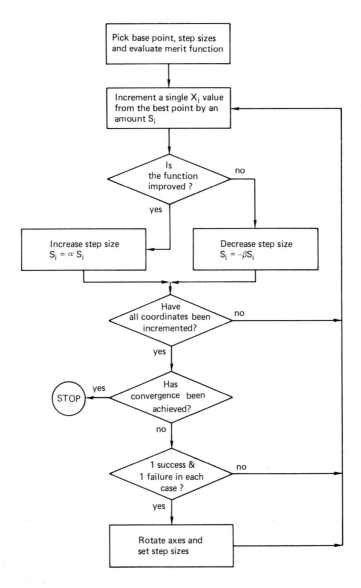

Figure 6-4 The Rosenbrock pattern search algorithm.

stepped a distance S_i parallel to the design variable axis, and the merit function is evaluated. As this process proceeds, if the value of F decreases, the move is termed a success and the step distance is increased by the formula

$$S_i = \alpha S_i$$

where α is a preselected value greater than 1. On the other hand, if the value of F

increases, the move is termed a failure, and the step distance is decreased by the formula

$$S_i = -\beta S_i$$

where β is a preselected factor less than 1. Once all the variables have been stepped, a convergence check is made. If the process has converged, the procedure terminates. If the process has not converged, an additional check is made to see if at least one success and one failure have occurred in each direction. If this combination of success and failure has not been achieved, the stepping procedure is repeated starting with the first variable. If at least one success and one failure in each direction have occurred, the axes are rotated so that the initial search direction is in a previously established direction of greatest improvement. Step sizes are then set, and the search continues in each of the variable directions using the new coordinate axes. Instead of moving a fixed step in each direction, this algorithm attempts to find the optimum point on each line. In so doing, the procedure continuously adjusts the step size for the pattern search. The combination of rotation of the ridge-following vector and the adjustment of the scale has made this algorithm an extremely powerful one for handling difficult optimization problems. Once the Rosenbrock search algorithm encounters a ridge, the direction of search becomes aligned with the ridge itself. This orientation is rather fortunate since for many engineering design problems the ridge will often point toward the extremum and the optimum will lie along the boundary line. For this reason, Rosenbrock's method will often overcome the problem posed in Figure 6-3 for the Hooke and Jeeves method. Unfortunately, some types of engineering problems have constraint boundaries made up of more than one curve or line. When two constraint boundaries intersect there will be a discontinuity in the behavior of the ridge. The Rosenbrock method may tend to terminate prematurely at this type of location. As before, this type of difficulty can be identified by trying different starting points to see if the termination is at the same location along the boundary. When this type of situation occurs, the designer could switch to a different type of penalty function that keeps the search somewhat away from the boundaries for the initial stages of the search. Another remedy to this problem would be to set the boundary as an equality constraint and then solve the reduced problem to see what optimum results. A third alternative for handling this situation is to try another algorithm, such as the simplex method. Before we describe such methods, we will consider an example application of the Rosenbrock search method.

EXAMPLE 6-2

Suppose that it is desired to solve Colville's problem posed in Example 6-1 using the Rosenbrock pattern search method. A BASIC program that implements this problem follows. This program uses the same penalty function that was used by the Hooke and Jeeves method in Example 6-1.

```
100   GOTO 9000
110 :
500   REM *********************
510   REM * MERIT FUNCTION FOR *
520   REM * COLVILLE'S MERIT   *
530   REM * SURFACE            *
540   REM *********************
550 :
560 F = 100 * (X(2) - X(1) ^ 2) ^ 2 + (1
  - X(1)) ^ 2 + 90 * (X(4) - X(3) ^ 2) ^
2 + (1 - X(3)) ^ 2 + 10.1 * ((X(2) - 1)
^ 2 + (X(4) - 1) ^ 2) + 19.8 * (X(2) - 1
) * (X(4) - 1)
570 :
580   REM   CHECK CONSTRAINTS
590   FOR IX = 1 TO NV
600   IF   ABS (X(IX)) > 10 GOTO 650
610   NEXT IX
620   RETURN
630 :
640   REM   PENALIZE
650 F = 20000!
660   RETURN
670 :
2000   REM *********************
2010   REM * THIS SUBROUTINE   *
2020   REM * APPLIES THE       *
2030   REM * ROSENBROCK PATTERN*
2040   REM * SEARCH ALGORITHM  *
2050   REM 0 TO FIND THE UNCON-*
2060   REM * STRAINED MINIMUM  *
2070   REM * OF A MERIT FUNC-  *
2080   REM * TION.  THE USER   *
2090   REM * MUST SUPPLY A     *
2100   REM * STARTING VECTOR   *
2110   REM * X(I).             *
2120   REM *                   *
2130   REM * PARAMETERS:       *
2140   REM *                   *
2150   REM *  NV  - THE NUMBER *
2160   REM *        OF DESIGN  *
2170   REM *        VARIABLES. *
2180   REM *                   *
2190   REM *                   *
2200   REM * 500  - SUBROUTINE *
2210   REM *        TO EVALUATE*
2220   REM *        THE MERIT  *
2230   REM *        VALUE "F"  *
2240   REM *                   *
2250   REM * EPS -  TEST VALUE *
2260   REM *        FOR MINIMUM*
2270   REM *        STEP SIZE  *
2280   REM *        TO STOP THE*
2290   REM *        SEARCH.    *
2300   REM *                   *
2310   REM *  ST  - STARTING   *
2320   REM *        STEP SIZE  *
2330   REM *                   *
2340   REM * ALPHA- SCALING    *
2350   REM *        FACTOR FOR *
2360   REM *        STEP SIZE  *
```

```
2370  REM *          INCREASE     *
2380  REM *          ALPHA>1      *
2390  REM *                       *
2400  REM * BETA - SCALING        *
2410  REM *          FACTOR FOR   *
2420  REM *          STEP SIZE    *
2430  REM *          REDUCTION    *
2440  REM *          0<BETA<1     *
2450  REM *                       *
2460  REM * FB  - ON RETURN       *
2470  REM *          THE BEST     *
2480  REM *          MERIT VALUE  *
2490  REM *          FOUND.       *
2500  REM *                       *
2510  REM * XB(I)- THE DESIGN     *
2520  REM *          VECTOR COR-  *
2530  REM *          RESPONDING   *
2540  REM *          TO FB.       *
2550  REM *                       *
2560  REM *********************
2570  :
2580  REM  SETUP
2590  GOSUB 500:FB = F
2600  FOR J = 1 TO NV
2610  FOR I = 1 TO NV
2620  M(I,J) = 0
2630  NEXT I
2640  M(J,J) = 1
2650  XB(J) = X(J)
2660  NEXT J
2670  :
2680  :
2690  REM  START STAGE HERE
2700  FOR I = 1 TO NV
2710  FAILURE(I) = 0:SUCCESS(I) = 0
2720  XE(I) = XB(I):FE = FB
2730  D(I) = 0:S(I) = ST
2740  NEXT I
2750  :
2760  FOR I = 1 TO NV
2770  FOR J = 1 TO NV
2780  X(J) = XE(J) + S(I) * M(I,J)
2790  NEXT J
2800  :
2810  GOSUB 500
2820  :
2830  IF F > FE GOTO 2940
2840  :
2850  SUCCESS(I)=1
2860  D(I) = D(I) + S(I)
2870  S(I) = S(I) * ALPHA
2880  FOR J = 1 TO NV
2890  XE(J)=X(J)
2900  NEXT J
2910  FE = F
2920  GOTO 2970
2930  :
2940  FAILURE(I) = 1
2950  S(I) =  - 1 * BETA * S(I)
2960  IF  ABS (S(I)) < EPS THEN  GOTO 3630
2970  NEXT I
2980  :
2990  :
3000  REM  ONE SUCCESS AND
```

```
3010   REM    ONE FAILURE IN
3020   REM    EACH DIRECTION?
3030   FOR I = 1 TO NV
3040   IF SUCCESS(I) = 0 GOTO 2760
3050   IF FAILURE(I) = 0 GOTO 2760
3060   NEXT I
3070   :
3080   REM   ZERO A(I,J) ARRAY
3090   FOR I = 1 TO NV
3100   FOR J = 1 TO NV
3110   A(I,J) = 0
3120   NEXT J
3130   NEXT I
3140   REM   ROTATE COORDINATES
3150   FOR I = 1 TO NV
3160   FOR J = 1 TO NV
3170   FOR K = I TO NV
3180   A(I,J) = D(K) * M(K,J) + A(I,J)
3190   NEXT K
3200   DD(I,J) = A(I,J)
3210   NEXT J
3220   NEXT I
3230   :
3240   DL(1) = 0
3250   FOR K = 1 TO NV
3260   DL(1) = DL(1) + DD(1,K) * DD(1,K)
3270   NEXT K
3280   DL(1) =  SQR (DL(1))
3290   FOR J = 1 TO NV
3300   M(1,J) = DD(1,J) / DL(1)
3310   NEXT J
3320   FOR I = 2 TO NV
3330   FOR J = 1 TO NV
3340   SA = 0
3350   FOR KK = 1 TO I - 1
3360   SB = 0
3370   FOR K = 1 TO NV
3380   SB = SB + A(I,K) * M(KK,K)
3390   NEXT K
3400   SA=SB*M(KK,J)+SA
3410   NEXT KK
3420   DD(I,J) = A(I,J) - SA
3430   NEXT J
3440   NEXT I
3450   FOR I=2 TO NV
3460   DL(I) = 0
3470   FOR K = 1 TO NV
3480   DL(I)=DL(I)+DD(I,K)*DD(I,K)
3490   NEXT K
3500   DL(I)=SQR(DL(I))
3510   FOR J=1 TO NV
3520   M(I,J)=DD(I,J)/DL(I)
3530   NEXT J
3540   NEXT I
3550   FOR I=1 TO NV
3560   XB(I)=XE(I)
3570   NEXT I
3580   FB=FE
3590   :
3600   GOTO 2690
3610   :
3620   REM   FINISH THE ANSWER
3630   FOR I=1 TO NV
3640   XB(I)=XE(I)
3650   NEXT I
```

```
3660 FB=FE
3670 RETURN
3680 :
9000   REM ********************
9010   REM * DRIVER PROGRAM    *
9020   REM * PERFORMS AN UNCON-*
9030   REM * STRAINED OPTIMIZA-*
9040   REM * TION OF A MERIT   *
9050   REM * FUNCTION.         *
9060   REM ********************
9070 :
9080 NV = 4:EPS = .00001
9090 BETA = .5:ALPHA = 2:ST = .001
9100   DIM XB(NV),X(NV),XE(NV)
9110   DIM D(NV),S(NV),A(NV,NV),M(NV,NV),
DD(NV,NV),DL(NV),SUCCESS(NV),FAILURE(NV)
9120 :
9130   REM  FEASIBLE START POINT
9140 X(1) =  - 3!:X(2) =  - 1:X(3) =  -
3:X(4) =  - 1
9150 :
9160   REM  CALL ROSENBROCK SEARCH
9170   GOSUB 2000
9180 :
9190 PRINT "FINAL RESULT IS":PRINT "F=";
FB:PRINT "X(1)=";XB(1):PRINT "X(2)=";XB(
2)
9200   PRINT "X(3)=";XB(3)
9210 PRINT "X(4)=";XB(4)
9220 END
```

To run this program the user must set the values of ALPHA, the step size acceleration factor; BETA, the step size deceleration factor; ST, the starting step size; and EPS, the factor used to measure convergence. The convergence test that is used in this particular problem is to terminate the search if the step size is decreased below the value of EPS during any stage. Since the step sizes are reset to the starting value ST for each step, this procedure is reasonably rigorous as a criterion to use. The starting value for this problem is the same point as that used for Example 6-1. It should be noted that the program requires quite a bit of extra working storage in the form of arrays in order to handle the rotation of coordinates. The DIM statements at lines 9100 and 9110 show this fact. The output of this program is as follows.

```
FINAL RESULT IS
F= 2.57582E-06
X(1)= 1.000642
X(2)= 1.00138
X(3)= .9993269
X(4)= .9986362
```

This output required 78 seconds to complete on an IBM PC. In performing this optimization, the program has made 267 merit evaluations. Although this performance would indicate that the algorithm is more efficient than the Hooke and Jeeves method, such a comparison is probably not fair since the algorithms are greatly different. Unlike the Hooke and Jeeves method, this algorithm is quite sensitive to the choices of ALPHA, BETA, and ST. Small changes in these parameters can greatly change the speed of convergence. For this reason the user would be well advised to monitor the performance of the algorithm as it proceeds by means of a print statement in the merit function subprogram. If it appears that the solution is not proceeding with reasonable speed toward convergence or if the solution that is approached is not the one desired, the process can be restarted with another set of initial parameters before significant wasted effort is expended.

6-3 SIMPLEX METHODS

To understand simplex methods, it is necessary to understand the concept of a simplex. A simplex is an n-dimensional, closed geometric figure in space that has straight-line edges intersecting at $n + 1$ vertices. In two dimensions this figure would be a triangle. In three dimensions it would be a tetrahedron. Search schemes based on the simplex utilize observations of the merit function at each of the vertices. The basic move in this method is a reflection to generate a new vertex point and thus a new simplex. The choice of reflection direction and the choice of the new vertex point depends on the location of the worst point in the simplex, as demonstrated in Figure 6-5. The new point is called the *complement* of the worst point. If the newest point in a new simplex has the worst value in that new simplex, the algorithm would oscillate back and forth rather than moving onward to the extremum. When this happens the basic move pattern is modified by using the second-worst point to locate the complement. The simplex technique tends to move the centroid of the area it bounds toward the extremum.

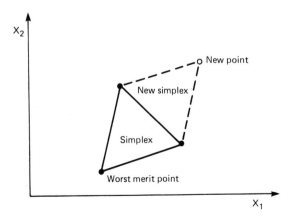

Figure 6-5   A simplex in two-dimensional space.

If, in addition to the reflection process, the edges of the simplex are allowed to contract and expand in size, the algorithm becomes that presented by Nelder and Mead. This general algorithm adapts to the local landscape in order to reach the minimum of a unimodal function of the form

$$\text{merit} = F(x_1, x_2, \ldots, x_n)$$

The logic diagram for this method is shown in Figure 6-6. The algorithm can be described as follows. First, the initial simplex is placed in the design space. The location of the $n + 1$ vertices of the initial simplex are defined in terms of a scale factor a and the number of design variables n. Special simplex parameters p and q are defined as

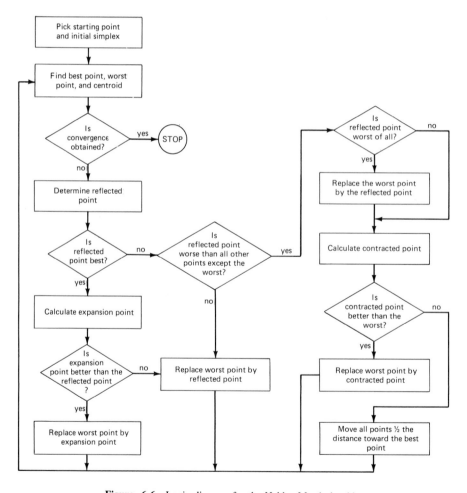

Figure 6-6 Logic diagram for the Nelder–Mead algorithm.

$$p = a\left(\frac{\sqrt{n+1} + n}{n\sqrt{2}}\right)$$

$$q = a\left(\frac{\sqrt{n+1} - 1}{n\sqrt{2}}\right)$$

Then the $n + 1$ vertices of the initial simplex will be related to the starting vector x_i by,

for j not equal to $i + 1$,

$$xs_{i,j} = x_i + q \qquad i = 1, 2, \ldots, n \quad \text{and} \quad j = 1, 2, \ldots, n + 1$$

and for $j = i + 1$,

$$xs_{i,j} = x_i + p \qquad i = 1, 2, \ldots, n \quad \text{and} \quad j = 1, 2, \ldots, n + 1$$

Once the simplex is formed, the merit function is evaluated at each vertex. One of the vertices will be the worst (highest for a minimum) point $xs_{(i,\ jw)}$. The basic operations used in the search method are defined using the centroid coordinates of all vertices in the simplex computed as a numerical average by using all vertices except the worst as follows:

$$xx_i = \frac{\text{sum } (xs_{i,j}) - x_{i,jw}}{n}$$

A reflected point is located using the relationship

$$xr_{i,j} = xx_i + \alpha(xx_i - xs_{i,jw})$$

where α is a positive constant called the reflection coefficient.

A contracted point is defined using the relationship

$$xc_{i,j} = xx_i - \beta(xx_i - xs_{i,jw})$$

where β is a preselected constant that is less than unity. An expansion point is defined using the relationship

$$xe_{i,j} = xx_i + \gamma(xr_{i,j} - xx_i)$$

where γ is a preselected constant called the expansion coefficient and is greater than unity.

As mentioned previously, the process of moving toward an optimum is accomplished by shrinking and expanding the simplex. This motion will be accomplished according to the following logic:

Step 1. The initial simplex is formed and the merit value FS_j is found at each vertex.

Step 2. The worst FW and best FB merit locations are isolated. The centroid vector xc_i and the merit value FX at the centroid are computed.

Step 3. The problem is tested for convergence using the criterion

$$\text{TEST} = [(\text{SUM}(\text{FS}_i - \text{FX})^2 - (\text{FW} - \text{FX})^2)/n]^{.5}$$

If TEST is smaller than a preset value, the process is terminated and the centroid is declared to be the location of the optimum.

Step 4. A reflection point and its corresponding merit value FR are calculated. Depending on the size of the merit value at this point, one of several alternative operations will take place.

Step 5. If the reflected point has a more favorable merit value than the best found so far, an expansion is done. The most favorable of either the expansion point or the reflected point is used to replace the worst merit vertex in the simplex and the process starting at step 2 is repeated.

Step 6. If the reflected point is worse than every point in the simplex except the worst FW, the reflected point is used to replace the worst point and the process continues with step 7.

Step 7. A contracted point vector is computed. If the contraction point is better than the worst point in the vertex, the contraction point is used to replace the worst point and the process is restarted at step 2. If the contraction point is not better than even the worst point in the simplex, the simplex is allowed to shrink by moving each of the simplex points one-half the distance toward the best point $xs_{i,jb}$ and the process is restarted at step 2. The shrinking formula used is

$$xs_{i,j}(\text{new}) = \frac{xs_{i,jb} + xs_{i,j}}{2}$$

The Nelder–Mead algorithm has the versatility to adapt itself to the local landscape of the merit surface. It will elongate down inclined planes, it will change direction on encountering a valley at an angle, and it will contract in the neighborhood of an extremum. The method is quite effective and is computationally compact. To illustrate the use of the Nelder–Mead algorithm, we will now undertake an example problem.

EXAMPLE 6-3

Suppose that it is desired to solve Colville's problem posed in Example 6-1 using the simplex method of Nelder and Mead. A BASIC program that implements this problem follows. This program uses the same penalty function that was used by the Hooke and Jeeves method in Examples 6-1 and 6-2.

```
100   GOTO 9000
110 :
500   REM ***********************
510   REM * MERIT FUNCTION FOR  *
520   REM * COLVILLE'S MERIT     *
530   REM * SURFACE              *
540   REM ***********************
550 :
560 F = 100 * (X(2) - X(1) ^ 2) ^ 2 + (1
 - X(1)) ^ 2 + 90 * (X(4) - X(3) ^ 2) ^
2 + (1 - X(3)) ^ 2 + 10.1 * ((X(2) - 1)
^ 2 + (X(4) - 1) ^ 2) + 19.8 * (X(2) - 1
) * (X(4) - 1)
570 :
580   REM  CHECK CONSTRAINTS
590   FOR IX = 1 TO NV
600   IF  ABS (X(IX)) > 10 GOTO 650
610   NEXT IX
620   RETURN
630 :
640   REM  PENALIZE
650 F = 20000!
660   RETURN
670 :
2000   REM *********************
2010   REM * THIS SUBROUTINE   *
2020   REM * APPLIES THE       *
2030   REM * NELDER MEAD       *
2040   REM * SEARCH ALGORITHM  *
2050   REM * TO FIND THE UNCON-*
2060   REM * STRAINED MINIMUM  *
2070   REM * OF A MERIT FUNC-  *
2080   REM * TION.  THE USER   *
2090   REM * MUST SUPPLY A     *
2100   REM * STARTING VECTOR   *
2110   REM * X(I).             *
2120   REM *                   *
2130   REM * PARAMETERS:       *
2140   REM *                   *
2150   REM *  NV  - THE NUMBER *
2160   REM *        OF DESIGN  *
2170   REM *        VARIABLES. *
2180   REM *                   *
2190   REM *                   *
2200   REM * 500  - SUBROUTINE *
2210   REM *        TO EVALUATE*
2220   REM *        THE MERIT  *
2230   REM *        VALUE "F"  *
2240   REM *                   *
2250   REM * EPS -  TEST VALUE *
2260   REM *        FOR MINIMUM*
2270   REM *        STEP SIZE  *
2280   REM *        TO STOP THE*
2290   REM *        SEARCH.    *
2300   REM *                   *
2310   REM * A    - SIDE LENGTH*
2320   REM *        OF SIMPLEX *
2330   REM *                   *
2340   REM * ALPHA- REFLECTION *
2350   REM *        COEFFICIENT*
2360   REM *        USUALLY=1.0*
2370   REM *                   *
2380   REM * BETA - CONTRACTION*
2390   REM *        COEFFICIENT*
```

```
2400  REM *          0<BETA<1     *
2410  REM *                       *
2420  REM * GAMMA- EXPANSION      *
2430  REM *          COEFFICIENT  *
2440  REM *          GAMMA>1.0    *
2450  REM *                       *
2460  REM *  FB  -  ON RETURN     *
2470  REM *          THE BEST     *
2480  REM *          MERIT VALUE  *
2490  REM *          FOUND.       *
2500  REM *                       *
2510  REM * XB(I)- THE DESIGN     *
2520  REM *         VECTOR COR-   *
2530  REM *         RESPONDING    *
2540  REM *         TO FB.        *
2550  REM *                       *
2560  REM ********************
2570  :
2580  REM  LOAD STARTING SIMPLEX
2590 P = A * ( SQR (NV + 1) + NV -1)/(NV
*SQR(2))
2600 Q = A * ( SQR (NV + 1) - 1) /(NV*SQ
R(2))
2610  FOR I = 1 TO NV
2620 XS(I,1) = X(I)
2630  NEXT I
2640  FOR J = 2 TO NV + 1
2650  FOR I = 1 TO NV
2660  IF (J = I + 1) GOTO 2690
2670 XS(I,J) = X(I) + Q
2680  GOTO 2700
2690 XS(I,J) = X(I) + P
2700  NEXT I
2710  NEXT J
2720 :
2730  REM  FIND FUNCTION
2740  REM  VALUES HERE
2750 FOR J=1 TO NV+1
2760 FOR I=1 TO NV
2770 X(I)=XS(I,J)
2780 NEXT I
2790 GOSUB 500:FS(J)=F
2800 NEXT J
2810 :
2820 REM FIND WORST, BEST
2830 JW=1:FW=FS(1)
2840 JB=1:FB=FS(1)
2850 FOR J=1 TO NV+1
2860 IF FS(J)<FB GOTO 2890
2870 IF FS(J)>FW GOTO 2910
2880 GOTO 2920
2890 JB=J:FB=FS(J)
2900 GOTO 2920
2910 JW=J:FW=FS(J)
2920 NEXT J
2930 :
2940  REM  FIND CENTROID
2950 FOR I=1 TO NV
2960 SUM=0
2970 FOR J=1 TO NV+1
2980 SUM=SUM+XS(I,J)
2990  NEXT J
3000 SUM = SUM - XS(I,JW)
3010 XX(I) = SUM / (NV)
3020 X(I) = XX(I)
```

```
3030   NEXT I
3040   GOSUB 500:FX = F
3050 :
3060   REM   CHECK FOR CONVERGENCE
3070 SUM = 0
3080   FOR J = 1 TO NV + 1
3090 SUM=SUM+(FS(J)-FX)^2
3100 NEXT J
3110 SUM = SUM - (FW - FX) ^ 2
3120 TEST =  SQR (SUM / NV)
3130   IF TEST < EPS GOTO 3910
3140 :
3150   REM   FIND REFLECTED POINT
3160   FOR I = 1 TO NV
3170 XR(I) = XX(I) + ALPHA * (XX(I)-XS(I
,JW))
3180 X(I) = XR(I)
3190   NEXT I
3200   GOSUB 500:FR = F
3210 :
3220   REM   IS REFLECTED BEST?
3230   IF FR > FB GOTO 3460
3240 :
3250   REM   CALCULATE EXPANSION PT.
3260   FOR I = 1 TO NV
3270 XE(I) = XX(I) + GAMMA * (XR(I) - XX
(I))
3280 X(I) = XE(I)
3290   NEXT I
3300   GOSUB 500:FE = F
3310 :
3320   REM   EXPANSION < REFLECTED?
3330   IF FE < FR GOTO 3400
3340 FW = FR:FS(JW) = FR
3350   FOR I = 1 TO NV
3360 XS(I,JW) = XR(I)
3370   NEXT I
3380   GOTO 2820
3390 :
3400 FW = FE:FS(JW) = FE
3410   FOR I = 1 TO NV
3420 XS(I,JW) = XE(I)
3430   NEXT I
3440   GOTO 2820
3450 :
3460   REM   IS FR WORSE THAN
3470   REM   ALL EXCEPT FW?
3480   REM   IF SO GO THROUGH
3490   REM   CONTRACTION
3500   REM   IF NOT REPLACE WORST
3510   REM   WITH FR.
3520   FOR J = 1 TO NV + 1
3530   IF J = JW GOTO 3560
3540   IF FR > FS(J) GOTO 3560
3550   GOTO 3580
3560   NEXT J
3570   GOTO 3710
3580 FW = FR:FS(JW) = FR
3590   FOR I = 1 TO NV
3600 XS(I,JW) = XR(I)
3610   NEXT I
3620   GOTO 2820
3630 :
3640 REM REFLECTED>WORST?
3650   IF FR < FW GOTO 3720
```

```
3660 FW = FR:FS(JW) = FR
3670   FOR I = 1 TO NV
3680 XS(I,JW) = XR(I)
3690   NEXT I
3700 :
3710   REM  CONTRACTED POINT
3720   FOR I = 1 TO NV
3730 XC(I) = XX(I) - BETA * (XX(I) - XS(
I,JW))
3740 X(I) = XC(I)
3750   NEXT I
3760   GOSUB 500:FC = F
3770 :
3780   IF FC < FW GOTO 3860
3790   FOR J = 1 TO NV + 1
3800   FOR I = 1 TO NV
3810 XS(I,J) = (XS(I,JB) + XS(I,J)) / 2
3820   NEXT I
3830   NEXT J
3840   GOTO 2730
3850 :
3860 FW = FC:FS(JW) = FC
3870   FOR I = 1 TO NV
3880 XS(I,JW) = XC(I)
3890   NEXT I
3900   GOTO 2820
3910   REM  FINISH ANSWER
3920 FB = FX
3930   FOR I = 1 TO NV
3940 XB(I) = XX(I)
3950   NEXT I
3960   RETURN
3970 :
9000   REM *****************
9010   REM * DRIVER PROGRAM *
9020   REM *****************
9030 :
9040 NV = 4:A = 9
9060 M = NV + 1
9070 ALPHA = 1:BETA = .5:GAMMA = 2:EPS =
 .0001
9080   DIM X(NV),XB(NV),XS(NV,M),FS(NV +
1),XX(NV),XR(NV),XC(NV),XE(NV)
9090 X(1) =  - 3:X(2) =  - 1:X(3)=  - 3:
X(4) =  - 1
9100 :
9110   GOSUB 2000
9120 :
9130   REM  PRINT RESULTS
9140   PRINT "FINAL RESULT IS": PRINT "F=
";FB
9150   FOR I=1 TO NV
9160 PRINT "X(";I;")=";XB(I)
9170   NEXT I
9180 END
```

To run this program the user must set the value of A, the simplex side length. For this problem we have chosen a value of 9.0 by experimentation with various possible values. The user must also choose a value for ALPHA, the reflection coefficient; BETA, the contraction coefficient; GAMMA, the expansion coefficient; and EPS, the factor used to measure convergence. The values used for these selections are those recommended by Nelder and Mead in their original paper describing the method. The starting value for this problem is the same point as that used for Example 6-2. It should be noted that the program requires quite a bit of extra working storage in the form of arrays in order to handle the simplex coordinates as well as the contraction, expansion, and reflection vectors. The DIM statement at line 9080 shows this fact. The output of this program is as follows:

```
FINAL RESULT IS
F= 4.919533E-06
X( 1 )= .9998472
X( 2 )= .9995398
X( 3 )= .9999677
X( 4 )= .9999936
```

This output required 103 seconds to complete on an IBM PC. In performing this optimization, the program has made 326 merit evaluations and performed 119 iterations through the Nelder–Mead cycle. Although this performance would indicate that the algorithm is less efficient than either of the previous two methods, such a comparison is probably not fair since the algorithms are greatly different. Thus for some types of problems this method would be expected to outperform either the Hooke and Jeeves method or the Rosenbrock method. As before, small changes in the input parameters can lead to a different performance from that presented here. In some cases the speed of convergence is too slow to be practical, and in other cases the convergence is not to the correct optimum. For this reason, the user would be well advised to monitor the performance of the algorithm as it proceeds by means of a print statement in the merit function subprogram. If it appears that the solution is not proceeding with reasonable speed toward convergence or if the solution that is approached is not the one desired, the process can be restarted with another set of initial parameters before significant wasted effort is expended.

6-4 CONSIDERATIONS IN THE SELECTION OF A PATTERN SEARCH METHOD

Pattern search methods are especially useful for multidimensional optimization problems when derivatives of the merit function are difficult or impossible to obtain. This does not mean that gradient methods should always be used if the derivatives can be found, since pattern search methods have considerable promise in speed and performance compared with other methods. Since they have the ability to follow ridges in the design space, they also have certain convergence advantages over gradient methods, which are difficult to use for these conditions. Because of their basic simplicity, pattern search methods are easily programmed and are considered by many experts to be the best general-purpose methods for multidimensional optimization. A few fundamental facts are useful to keep in mind when applying pattern search methods. These involve the selection of a particular pattern search algorithm, the implementation of a pattern search program, and the modification of the methods to enhance their utility. Let us look briefly at each of these topics.

Selection of a Particular Algorithm

The selection of a particular pattern search method will depend on the nature of the design space. Since the Hooke and Jeeves is the easiest method to program, it is frequently the first choice for optimization problems. If this method exhibits the characteristic of being stalled at a boundary ridge, the Rosenbrock or the Nelder–Mead method may be a better choice. Since each of the methods presented in this chapter is based on an entirely different procedure, they tend to have rather unique performance characteristics. Thus one method may work well for a particular problem even if the other two are poorly suited. It becomes desirable, then, for the designer to be able to interchange the various algorithms without the need to modify his or her driver program or merit subroutine. The example programs in this chapter fit this requirement. By experimenting with several different techniques, the designer can usually expand the probability of achieving a successful solution.

Implementation of Pattern Search Programs

In implementing a pattern search algorithm it is well to include some print statements in the merit function subroutine in order to monitor the performance of the algorithm as it progresses. If the user sees that the method is not making progress, the program may be stopped before significant time is wasted. This is, of course, one of the advantages of using the small computer over the larger computers. In any given method there are a number of parameters that need to be preselected by the user. Example choices of these parameters are presented in the example problems of this chapter. These will frequently be good selections for a first attempt at optimization using a given pattern search method. If these choices do not succeed in finding a solution, other choices may be tried as long as the new choices do not violate the basic requirements of these variables. The choice of initial starting values for the design variables is an extremely important matter. Often a change in these values can mean the difference between no solution and a quick solution. Also, the use of different starting values can be a useful tool to find a global optimum rather than a local optimum for a multimodal merit surface. The selection of a particular type of penalty function can also enhance the performance of an optimization algorithm when the solution lies along a boundary. If the method appears to be stalled at a boundary, the problem can sometimes be handled by changing the nature of the penalty surface to avoid design choices exactly on the boundary until the search progress has gone on for some time. A review of Chapter 2 will help the user to find alternative formulations for penalty surfaces.

Modification of Pattern Search Methods for Enhanced Utility

In this chapter it was suggested that the methods be modified by combining two or more search schemes into a single optimization subroutine. Typical examples of this involve the combination of random search with a Hooke and Jeeves search or the use of a method which alternates between the use of a simplex search and a Rosenbrock search. The rationale for this approach is quite sound. If one of the two methods is successful and the other is not, the user will still find a solution. Thus the user can double the likelihood of success through the use of such "hybrid" methods. Eason has done an extensive study of various methods and suggests that hybrid methods and methods with interchangeable subroutines are generally superior to single-algorithm methods. The user may wish to experiment with the methods presented so far in this book to see what types of hybrid methods can be assembled and how they perform relative to single-algorithm methods.

PROBLEMS

6-1 In Example 6-1 the starting values of $(-3, -1, -3, -1)$ converged to the global optimum at $(1, 1, 1, 1)$. Experiment with this algorithm to see what alternative starting points will lead to the local optimum at $(-1, 1, -1, 1)$.

6-2 In Example 6-2 it was pointed out that the choice of ALPHA, BETA, and ST can have a significant influence on the speed of convergence and on whether the method converges to a local optimum or a global optimum. Undertake a parameter study for this problem to compare:
(a) Convergence time versus ALPHA.
(b) Convergence time versus BETA.
(c) Convergence time versus ST.

6-3 A well-known problem in optimization literature is known as the Rosenbrock post office parcel problem. It involves the physical design of a rectangular box to maximize the volume (i.e., to minimize $-V$):

$$F = \frac{-x_1 x_2 x_3}{1000}$$

subject to post office restrictions on the length plus girth and individual limits on the length, depth, and height:

$$0 \leq x_1 + 2(x_2 + x_3) \leq 72$$

$$0 \leq x_1 \leq 20$$

$$0 \leq x_2 \leq 11$$

$$0 \leq x_3 \leq 42$$

If one uses a feasible starting point of $(10, 10, 10, 10)$ with $F = -1$ and a simple penalty function similar to that found in Example 6-1, both the Hooke and Jeeves method and the Rosenbrock method will stall at the intersection of two of the constraint boundaries.

(a) Demonstrate this performance using the penalty function $F = -1$.

(b) Prepare a different penalty function that will not permit the solution to lie on the boundary (see Chapter 2) and see if you can find the global optimum at (20, 11, 15).

6-4 Solve Problem 6-3 using Rosenbrock's method.

6-5 In Section 6-1 it was proposed that the algorithm be modified to include a test for convergence based on a random search in the vicinity of the best point found. Modify the Hooke and Jeeves routine in Example 6-1 to accommodate this feature and compare the performance of your new subroutine with that used for the example.

6-6 The formula for the inductance of an air-core inductance coil having a single layer of wire is

$$L = \frac{0.2 \, a^2 N^2}{3a + 9b}$$

where N = number of turns of wire

         L = inductance (microhenrys)

         a = average diameter of the coil (inches)

         b = length of winding (inches)

It is desired to design a 60 μH coil that will take a minimum length of wire subject to

$$1.0 \le a \le 2.0$$

$$0.25 \le b \le 1.0$$

Find an optimum design for this coil by specifying the values of a and N.

6-7 In Example 6-3 the user was required to specify the initial length of the side of the simplex A. Undertake a parameter study for the problem solved in Example 6-3 to compare the convergence time to different choices for A. What happens to the performance as the value of A is changed from 9.0?

6-8 In Section 6-4 it was suggested that an optimization technique could be developed that would alternate between a simplex method and the Rosenbrock search scheme. See if you can modify the subroutines in Examples 6-2 and 6-3 to develop a scheme that will alternate between these two methods after having done 20 iterations in each. How does your new program compare in speed of convergence to the performance of the single-algorithm method?

6-9 In Section 6-4 it was pointed out that pattern search methods are often useful in handling problems that could have been solved by a gradient-based method. Run Example 5-1 using the Hooke and Jeeves method and compare your results with the gradient-based methods of Chapter 5.

6-10 The design of a three-bar planar truss is a classic problem in optimization. [See, for example, the work of Menkes, Sandor, and Wang (1972).] For this truss the designer wants to minimize the total weight of the structure:

$$F = rl_1d_1 + rl_2d_2 + rl_3d_3$$

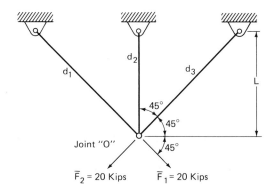

d_i , i = 1, 2, 3   cross-section areas of bars, in.^{2}

F_k , k = 1, 2   external forces

L   length of bar 2, in.

Symmetric planar three bar truss

where   r = weight density
        l_i = length of the *i*th member
        d_i = cross-sectional area of the *i*th member

The constraints for this problem are that no cross-sectional area can be equal to or smaller than zero and that the tensile stress and the compressive stress not exceed the allowable values of 20 ksi for tensile strength and 15 ksi for compressive strength. This leads to the following six constraint equations:

$$1 + \frac{S_{11}(d_i)}{15} \geq 0$$

$$1 + \frac{S_{21}(d_i)}{15} \geq 0$$

$$1 + \frac{S_{31}(d_i)}{15} \geq 0$$

$$1 - \frac{S_{11}(d_i)}{20} \geq 0$$

$$1 - \frac{S_{21}(d_i)}{20} \geq 0$$

$$1 - \frac{S_{31}(d_i)}{20} \geq 0$$

where

$$S_{11} = 20d_1^{-1} - \frac{20d_2}{D}$$

$$S_{21} = \frac{20(2)^{0.5}d_1}{D}$$

$$S_{31} = \frac{-20d_2}{D}$$

with $D = (2)^{0.5}d_1^2 + 2d_1 d_2$. Owing to symmetry, one can reduce the number of dimensions in the problem by letting $d_1 = d_3$. Using unit values for the length L and for the density, start with the design point (1.0, 1.0) and see if you can find an optimum for this problem.

REFERENCES

1. Eason, E. D., and Fenton, R. G., A Comparison of Numerical Optimization Methods for Engineering Design, *ASME Paper 73-DET-17,* 1974.
2. Eason, E. D., and Fenton, R. G., Testing and Evaluation of Numerical Methods for Design Optimization, *UTME-TP 7204,* University of Toronto, Sept. 1972.
3. Hooke, R., and Jeeves, T. A., "Direct Search Solution of Numerical and Statistical Problems," *Journal of the Association for Computing Machinery,* Vol. 8, 1961, pp. 212–229.
4. Menkes E. G., Sandor, G. N., and Wang, L. R., "Optimum Design of Composite Multilayer Shell Structures, *ASME Paper 71-WA/DE 12,* 1971.
5. Nelder, J. A., and Mead, R., "A Simplex Method for Function Minimization," *Computer Journal,* Vol. 7, 1964, pp. 308–313.
6. Shoup, T. E., *A Practical Guide to Computer Methods for Engineers,* Prentice-Hall, Inc., Englewood Cliffs, N.J., 1979.

7

ADVANCED TOPICS
IN OPTIMIZATION

There is no substitute for the use of human judgment in the design process. (Courtesy of Manufacturing and Consulting Services, Inc.)

Real-world design problems frequently have a number of distinguishing characteristics that add to the complexity of their formulation. Among these characteristics are the following:

1. They may be multidisciplinary with many design variables and many constraints.
2. They may have special types of constraints that lend them to particular types of solution techniques (e.g., they may have constraints and merit functions that are linear relationships of the design variables).
3. They may have multiple measures of merit that are often in conflict with each other and thus call for a compromise (e.g., they may require the simultaneous minimization of cost, minimization of weight, and the maximization of strength).
4. They may contain information that is based on judgment rather than scientific fact or principle (e.g., they may contain decisions that are made on esthetic considerations or on market appeal).
5. They may contain the need for information that is not available at the time of the problem analysis (e.g., they may require information about the future cost of inventory or material).
6. They may contain design variables that are able to have only discrete values (e.g., the dimensional size of a bolt or the number of passengers on a mass-transit train car).

It seems logical to ask how these real-world problems in optimization can be handled since the designer is likely to encounter them in actual practice. It is the purpose of this chapter to discuss how some of these issues can be handled by means of some of the techniques described earlier. It is also the purpose of this chapter to describe briefly the availability of special, advanced methods to help with these topics. Although it is beyond the intended scope of this book to describe these methods in great detail, the material will present information from the engineering literature to assist the interested reader in probing deeper into these timely topics. A more thorough discussion of the advanced topics introduced here will be presented in the planned sequel to this work.

In considering the advanced topics of optimization it is well to maintain the perspective that the function of optimization in design is to provide support for human judgment. Recent literature in the field of optimization refers to this process as the *decision-support problem* and provides insight into the formulation of such design tasks. In spite of the sophistication of numerical procedures to assist with the optimization process, there is, to date, no adequate substitute for the human mind for the final decision process associated with design.

7-1 LINEAR PROGRAMMING IN OPTIMIZATION DESIGN

Linear programming is a special class of optimization problems in which the objective function is always expressed as a linear relationship of the design variables and the constraints are expressed as linear equalities or inequalities. Although the linear restriction may seem to be a significant limitation, a very large number of real-world problems fit into this category. Among these are problems associated with transportation scheduling, inventory maintenance, and investment management.

The problem to be solved is to minimize the function

$$F = \sum_{i=1}^{n} c_i x_i$$

subject to the linear constraints

$$\sum_{i=1}^{n} a_{mi} x_i \le b_m \qquad m = 1, 2, \ldots, M$$

Because of the special nature of this type of problem, it can be shown that the optimum solution must lie along one of the boundaries. In fact, the solution will usually lie at the intersection of one or more of the boundary lines. With this special characteristic, the search can be confined to the boundaries, and as a result, the solution is far easier to solve than if every point in the design space is considered by one of the methods presented earlier in this book. If there are only two design variables in a linear programming problem, the result can frequently be found by a graphical solution. Techniques for problems with more than two design variables are based on direct search procedures in which the boundaries are systematically examined. The best known approach is that developed by Dantzig in 1947 (Dantzig, 1963) and has been used extensively since that time.

If the linear restriction for the merit function is modified to allow for other types of formulations, the problem becomes that known as *nonlinear programming*. One subset of this field is known as *quadratic programming*. The main difference in this type of problem and the linear programming problem is that it is possible for an extreme value to exist at a point where the constraints are not present. This situation is often referred to as a *field optimum*. One of the better known approaches to the quadratic programming problem has been proposed by Wolfe in 1959. It is based on the use of a modified form of the simplex method.

A very powerful approach to utilizing the concepts of linear programming to handle nonlinear problems of large complexity is based on a linearization of merit functions utilizing either only first-order terms or both first-order and second-order terms of a Taylor's series expansion of the merit function and the nonlinear constraints. For a discussion of this topic, the interested reader should consult the work of Mistree, Hughes, and Phuoc (1981). This method solves a sequence of linearized problems of increasing accuracy and is called the method of *sequential linear programming*. In theory the linearization of a nonlinear design space removes the ability to find a field optimum (i.e., an optimum not on a boundary). Fortunately, this situation hardly ever occurs in complex, highly constrained systems. The work of Mistree et al. shows how a technique of constraint accumulation can be used to find a field optimum and thus eliminate this potential shortcoming of the method.

7-2 COMPROMISE PROBLEMS IN DESIGN

In many engineering design situations there will be problems that are framed in terms of a desire to maximize or minimize more than one quantity. For example, in the design of an airplane, it is desired to maximize strength while minimizing both weight and cost. None of the techniques described so far in this book can handle this type of problem in its basic form. What is required is a method for incorporating each of the merit functions into a single composite merit function. For example, if it is desired to minimize three different merit functions,

$$f_i \quad i = 1, 2, 3$$

a composite merit function can be assembled of the form

$$F = c_1 f_1 + c_2 f_2 + c_3 f_3$$

In this expression the c values are known as weighting factors and the overall merit function F is known as a trade-off function. The designer will need to pick the relative size of the weighting factors to reflect the relative importance of the various merit factors in the merit function. For example, if the first two merit factors are more important than the third, a possible choice might be

$$c_1 = 0.5 \quad c_2 = 0.5 \quad c_3 = 0.1$$

In this example, the relative importance of the first two factors is five times more significant than the third in its contribution to the trade-off function. As can be seen from this simple example, such a formulation allows the designer to specify the relative importance of the component parts of the composite merit function. Once a composite function has been assembled, the designer can employ any of the methods presented previously in this book to perform the optimization process. Once a solution has been found, the designer may wish to modify the weighting factors and rework the problem. There are, of course, special techniques in the engineering literature to assist the designer with the handling of compromise problems.

   The compromise decision-support problem has been studied extensively by Kuppuraju and Mistree (1983). In their work, Kuppuraju and Mistree have shown how to reduce the size of the problem significantly by eliminating redundant constraints before attempting to solve the problem. Figure 7-1 presents a broad classification scheme for compromise decision-support problems. For further information about this approach, the interested reader should consult the work of Kuppuraju and Mistree.

7-3 DISCRETE-VALUED PROBLEMS

Some problems in engineering design require that the design variables take on discrete values. These values may be integers or may be discrete real values, as would be the case for dimensional sizes for standard parts. When such problems occur, the designer is faced with an unusual difficulty since most optimization methods rely on the availability of an infinite range of choices for the design variables.

One approach to handling the discrete-valued problem involves the use of a real variable for the design value. Using this technique, the designer solves the optimization problem as if the design variable could have any real value. Once the optimum is found, the allowable discrete values within the vicinity of the optimum are tried to see if they represent reasonable choices that identify an optimum for the discrete problem. This procedure often involves reevaluation of the merit function for several different design values obtained by "rounding up" and "rounding down" from the values discovered by the traditional search.

Another approach to handling discrete design variables is that of using a variation of the total search scheme presented in Chapter 3. Since the restriction of discrete values limits the number of possible design choices to a finite list, the problem can often be formulated in such a way that all possible designs can be evaluated in order to find the best. In this sense the discrete-valued restriction can actually be used to an advantage to reduce the amount of effort required to discover a solution. This approach does have the added advantage of being able to handle nonunimodal

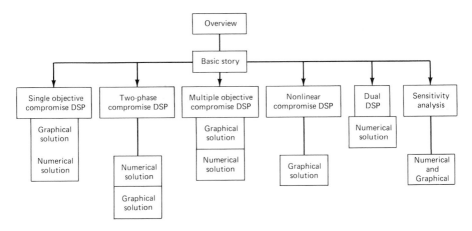

Figure 7-1 A classification scheme for compromise decision-support problems.

merit surfaces. As the complexity of the design problem increases or as the mix of discrete- and real-valued design variables increases, this approach soon becomes impractical. When this situation occurs, the random search approach presented in Chapter 4 can often prove to be helpful.

7-4 PREDICTION AND JUDGMENT IN OPTIMIZATION

Some types of problems in engineering design require the introduction of judgment about aesthetic factors or about future trends of a market or cost. Such types of variables in an optimization problem make the use of traditional search schemes rather difficult. It should be noted that the personal computer is probably the most useful tool available for implementing interactive prediction techniques since the invention of writing paper (Figure 7-2). Although it is possible to find optimization answers based on quantitative predictions of the future, in some cases the resulting designs are rather poor choices if the basic assumptions used to find the solution are not accurate. When this approach is used, it is wise to perform a sensitivity analysis to determine how much a given uncertainty in the input parameters will influence variations in the output results. If the results are relatively insensitive to the input assumptions, the designer can feel reasonably confident that the optimum design

Figure 7-2   The personal computer is an extremely powerful and useful tool for the implementation of interactive prediction techniques. (Courtesy of IBM.)

found has some meaning. If, on the other hand, the output is quite sensitive to the input assumptions, the results are usually of much less value.

Some types of engineering problems have parameters that are difficult to quantify under any set of assumptions. A good example of this is the market appeal of a given design choice. Even this nebulous concept can be quantified by means of market surveys. Thus a value from zero to 1 could be assigned to represent the results of an experimental determination of market appeal. In this way a qualitative measure can be converted to a real quantity that can be included in an optimization search scheme. Such an approach will often require interpolation of the experimental data. When this approach is used, it is wise to conduct a sensitivity analysis to see what effect the overall uncertainty of the input assumptions has on the output result.

REFERENCES

1. Dantzig, G. E., *Linear Programming and Extensions,* Princeton University Press, Princeton, N.J., 1963.
2. Kuppuraju, N., and Mistree, F., "Compromise: An Effective Approach to Solve Multiobjective Structural Design Problems," *Computers and Structures,* August 1984.
3. Mistree, F., "Design Damage of Tolerant Structural Systems," *Engineering Optimization,* Vol. 6, 1983, pp. 141–144.
4. Mistree, F., Hughes, O. F., and Phuoc, H. B., "An Optimization Method for the Design of Large, Highly Constrained Complex Systems," *Engineering Optimization,* Vol. 5, 1981, pp. 179–197.
5. Wilde, D. J., and Beightler, C. S., *Foundations of Optimization,* Prentice-Hall, Inc., Englewood Cliffs, N.J., 1967.
6. Wolfe, P., "The Simplex Method for Quadratic Programming," *Econometrica,* Vol. 27, 1959, pp. 382–398.

Appendix

GLOSSARY OF COMPUTER TERMS

ACCESS TIME — The time required to read data from/to input/output device.

ACCUMULATOR — A register that stores the results of a computer operation.

ADDER — A logic circuit that adds binary numbers.

ADDRESS — A binary number that identifies a specific memory storage location.

ALGORITHM — A computational procedure for solving a problem.

ALPHANUMERIC — A collection of characters containing both letters of the alphabet and numbers.

ANALOG-TO-DIGITAL CONVERTER — A device that converts external signals to a form the computer can recognize.

APL — A Programming Language. A higher-level terminal-oriented programming language.

APPLICATIONS PROGRAM — A program that solves a specific problem, such as inventory control or machine control.

ARITHMETIC-LOGIC UNIT — A logic circuit that performs both arithmetic and logical operations in a digital computer.

ARRAY — A list or table (matrix) of data.

ASCII — American Standard Code for Information Interchange. A binary code that represents alphanumeric characters and various symbols.

ASSEMBLER — A computer program that automatically converts assembly language mnemonics into machine language.

ASSEMBLY LANGUAGE — The next step above machine language. Substitutes easily remembered mnemonics for binary machine language instructions.

ASYNCHRONOUS — A computer operation that takes place whenever input information appears. The basic RS flip-flop, for example, is an asynchronous circuit.

AUXILIARY STORAGE — A storage that supplements the internal storage of the central processing unit. Also called secondary storage.

BACKGROUND PROCESSING — Processing of a low-priority program that takes place only when no higher-priority or real-time processing function is present.

BASE — The radix of a number system.

BASIC — Beginner's All-purpose Symbolic Instruction Code. A computer language used with most personal computers.

BATCH PROCESSING — A technique by which items to be processed must be coded and collected into groups prior to processing.

BINARY — A number system with the base 2. Also, a term used to describe a condition or electronic circuit which has only two states, usually on and off.

BINARY-CODED DECIMAL (BCD) — A number system used in digital computers and calculators that assigns a binary number to each of the 10 decimal digits.

BINARY DIGIT — The binary digits 0 or 1.

BISTABLE — An electronic circuit or device that has two operating states, such as a mechanical switch, indicator lamp, or flip-flop.

BIT — An abbreviation for binary digit.

BIT RATE — The rate at which binary digits, or pulse representations, appear on communications lines or channels.

BLOCK — A set of things, such as digits, characters, or words, handled as a unit.

BRANCH — A computer program procedure that transfers control from one instruction to another instruction elsewhere in the program.

BUFFER — A circuit that isolates one circuit from another circuit.

BUG — An error. It can be a mistake in a computer program or a defect in the operation of a computer.

BUS — One or more electrical conductors that transmit power or binary data to the various sections of a computer.

BYTE — A group of (usually) eight binary bits.

CALCULATOR — A microprocessor-based instrument designed primarily for solving mathematical problems.

CARD READER — A computer input mechanism that reads out the information contained on a punched card.

CARRIAGE — A control mechanism for a typewriter or printer that automatically feeds, skips, spaces, or ejects paper forms.

CASSETTE UNIT — A magnetic tape recorder that uses cassette tapes for storage. Widely used with microcomputers.

CENTRAL PROCESSING UNIT (CPU) — The arithmetic-logic unit and control sections of a digital computer.

CHARACTER — Any letter, number, or symbol that a digital computer can understand, store, or process.

CHIP — A thin slice of silicon up to a few tenths of an inch square with an integrated circuit containing from dozens to thousands of electronic parts on its surface.

CIRCUIT — A collection of electronic parts and electrical conductors that performs some useful operation.

CLOCK — A circuit that produces a sequence of regularly spaced electrical pulses to synchronize the operation of the various circuits in a digital computer.

COBOL — COmmon Business-Oriented Language. A higher-level programming language developed for programming business problems.

CODE — A method of representing letters, numbers, symbols, and data with binary numbers.

CODING — The process of translating problem logic represented by a flowchart into computer instructions and data.

CODING FORM — A form on which the instructions for programming a computer are written. Also called a coding sheet.

COMPATIBLE — A term applied to a computer system which implies that it is capable of handling programs devised for some other type of computer system.

COMPILER — A computer program that translates a high-level source-language program into machine-language programs suitable for execution on a particular computing system.

COMPUTER — An electronic device that processes discrete (digital) or approximate (analog) data.

COMPUTER SCIENCE — The field of knowledge that involves the design and use of computer equipment, including software development.

COMPUTER SYSTEM — A central processing unit together with one or more peripheral devices.

CONSOLE — That part of a computer used for communications between the computer operator or maintenance engineer and the computer.

CONTROL PANEL — A panel with input switches and output indicators that allows control of a computer system.

CONTROL SECTION — The electronic nerve center of a digital computer; the circuits that decode incoming instructions and activate the various sections of the computer in perfect synchronization. Part of the central processing unit (CPU).

CONTROL UNIT — The portion of the central processing unit that directs the step-by-step operation of the entire computing system.

CONVERSATIONAL MODE — A mode of operation where a user is in direct contact with a computer, and interaction is possible between human and machine without the user being conscious of any language or communications barrier.

COUNTER — A string of flip-flops that counts in binary.

CPU — The central processing unit.

CRT — Cathode-ray tube. The video display tube used in television sets and many computer terminals.

CYCLE — A specific time interval during which a regular sequence of computer events takes place.

DATA — Numbers, facts, information, results, signals, and almost anything else that can be fed into and processed by a computer.

DATA BASE — A comprehensive data file containing information in a format applicable to a user's needs and available when needed.

DATA COLLECTION — The gathering of source data to be entered into a computer system.

DATA PROCESSING — A term used in reference to operations performed by data processing equipment.

DATA PROCESSING CENTER — An installation of computer equipment which provides computing services for users.

DATA STRUCTURES — Arrangement of data (i.e., arrays, files, etc.).

DEBUG — The process of finding and fixing an error in a computer program or in the actual design of a computer.

DECIMAL — A number system with the base 10.

DECISION — A computer operation that compares two binary words or checks the status of a single bit or word and then takes a specified course of action.

DECODER — A combinational circuit that converts binary data into some other number system.

DECREMENT — To decrease the value of a number by a fixed value, often 1.

DEMULTIPLEXER — A combinational circuit that applies the logic state of a single input to one of several outputs.

DIAGNOSTICS — Statements printed by an assembler or compiler indicating mistakes detected in a source program.

DIGIT — A character in a number system that represents a specific quantity.

DIGITAL — Pertaining to the utilization of discrete integral numbers in a given base to represent all the quantities that occur in a problem or a calculation.

DIGITAL COMPUTER — A computer that uses discrete signals to represent numerical quantities. Nearly all modern digital computers are two-state, binary machines.

DIGITAL PLOTTER — An output unit that graphs data by an automatically controlled pen, plotted as a series of incremental steps.

DIGITAL-TO-ANALOG CONVERTER — A device that converts computer digital data into analog signals.

DIRECT-ACCESS STORAGE — Pertaining to the process of obtaining data from or placing data into storage where the time required for such access is independent of the location of the data most recently obtained or placed in storage. Also called random-access storage.

DISK MEMORY — *See* Magnetic disk memory.

DISK OPERATING SYSTEM — Software used to manage disk files and programs and to develop application software. Abbreviated DOS.

DISK PACK — The vertical stacking of a series of magnetic disks in a removable self-contained unit.

DISKETTE — *See* Floppy disk.

DISPLAY UNIT — A device that provides a visual representation of data. *See also* CRT.

DOCUMENTATION — An important part of computer design and program development. The process of recording in organized format a detailed list of operational or programming considerations.

DOWNTIME — The total elapsed time that a computer system is unusable because of a malfunction.

DUMP — Printing all or part of the contents of a storage device.

EBCDIC — Extended Binary Coded Decimal Interchange Code. An eight-bit code used for data representation in several computers.

EMULATE — To imitate one system with another, such that the imitating system accepts the same data, executes the same programs, and achieves the same results as the imitated system.

ENCODER — A combinational circuit that converts data from some other number system into binary.

EPROM — Erasable programmable read-only memory. A read-only storage device that can be erased to change its contents.

ERASE — To clear or remove data from a memory.

EXECUTE — To comply with or act on an instruction in a digital computer program.

FIELD — A particular category or grouping of data or instructions.

FILE — An organized collection of related data. For example, the entire set of inventory master data records makes up the inventory master file.

FIXED WORD — The condition in which a machine word always contains a fixed number of bits, characters, bytes, or digits.

FLIP-FLOP — The basic sequential logic circuit. A circuit that is always in one of two possible states.

FLOPPY DISK — A flexible disk (diskette) of oxide-coated Mylar that is stored in a plastic envelope. The entire envelope is inserted in the disk unit. Floppy disks are low-cost storage units and are widely used with microcomputers and minicomputers.

FLOWCHART — A diagram that shows the major steps or operations that take place in a computer program.

FOREGROUND PROCESSING — The automatic execution of the computer programs that have been designed to preempt the use of the computing facilities.

FORTRAN — FORmula TRANslation. A higher-level programming language designed for programming scientific-type problems.

GARBAGE — A term often used to describe incorrect answers from a computer program, usually resulting from equipment malfunction or a mistake in a computer program.

GATE — The simplest electronic logic circuit. A single gate may invert the logic state at its input or make a simple decision about the status of two or more inputs.

GENERAL-PURPOSE COMPUTER — A computer that is designed to solve a wide class of problems. The majority of digital computers are of this type.

GIGO — Garbage in, garbage out. A term used to describe the data into and out of a computer system [i.e., if the input data are bad (garbage in), the output data will also be bad (garbage out)].

HARD COPY — A paper printout of computer results or data.

HARDWARE — The electronic circuits in a computer.

HEXADECIMAL — A number system with the base 16. "Hex" numbers are convenient for representing 4-bit binary groups.

HIGHER-LEVEL LANGUAGE — A computer programming language that is intended to be independent of a particular computer.

HOUSEKEEPING — Operations that take place in a computer or a computer program that clear memories, check status registers, organize data, and otherwise set things up in preparation for a data processing operation.

ILLEGAL OPERATION — A program instruction that a computer cannot perform.

INCREMENT — To increase the value of a number by a fixed value, often 1.

INFORMATION — Data that have been organized into a meaningful sequence.

INFORMATION RETRIEVAL — A technique of classifying and indexing useful data in mass storage devices, in a format amenable to interaction with the user(s).

INPUT — The introduction of data from an external source into the internal storage unit of a computer.

INPUT/OUTPUT — A general term for the peripheral devices used to communicate with a digital computer and the data involved in the communication.

INPUT UNIT — A device used to transmit data into a central processing unit.

INSTRUCTION — A set of characters used to define a basic operation and to tell the computer where to find the data needed to carry out an operation.

INSTRUCTION REPERTOIRE — The complete set of machine instructions for a computer.

INTEGER — A whole number which may be positive, negative, or zero. It does not have a fractional part. Examples of integers are 526. -378, or 0.

INTEGRATED CIRCUIT — An electronic circuit formed on the surface of a tiny silicon chip.

INTELLIGENT TERMINAL — An input/output device in which a number of computer processing characteristics are physically built into the terminal unit.

INTERACTIVE — Highly communicative between the user and the computer system.

INTERFACE — A common boundary between two pieces of hardware or between two systems.

INTERNAL STORAGE — Addressable storage directly controlled by the central processing unit of a digital computer. It is an integral part of the central processing unit.

INTERPRETER — A computer program that translates and then executes a computer program a step at a time.

INTERRUPT — To disrupt temporarily the normal execution of a program by a special signal.

ITERATIVE PROCESS — A process in which the same procedure is repeated many times until the desired answer is produced.

JOB — A specified group of tasks prescribed as a unit of work for a computer.

K — A shorthand way of expressing the capacity of a computer memory. Corresponds to 2^{10} (1024). Therefore, 4K equals 4096.

KEYBOARD — A typewriterlike array of switches used to feed data into a digital computer manually.

LANGUAGE — The symbols, phrases, characters, and numbers used to communicate with a digital computer.

LIGHT PEN — A stylus used with CRT display devices to add, modify, and delete information on the face of the screen.

LINE PRINTER — A printer that prints a complete line of type in one operation.

LOGIC CIRCUIT — A gate or other circuit that responds to two-state signals.

LOOP — A sequence of computer instructions that is repeated one or more times until a desired result is achieved.

LSI — Large scale integration. Logic used in microprocessors, microcomputers, and other computers.

MACHINE ADDRESS — An address that is permanently assigned by the machine designer to a storage location. Also called absolute address.

MACHINE INDEPENDENT — A term used to indicate that a program is developed in terms of the problem rather than in terms of the characteristics of the computer system.

MACHINE LANGUAGE — Fundamental language of a computer. Programs written in machine language require no further interpretation.

MACROINSTRUCTION — A computer instruction composed of a sequence of microinstructions.

MAGNETIC CORE — A tiny ring of material that can store a single binary bit.

MAGNETIC DISK MEMORY — A memory system that stores and retrieves binary data on recordlike metal or plastic disks coated with a magnetic material.

MAGNETIC TAPE MEMORY — A memory system that stores and retrieves binary data on magnetic recording tape.

MAGNETIC TAPE UNIT — A device used to read and write data in the form of magnetic spots on reels of tape coated with a magnetizable material.

MAIN FRAME — The part of the computer that contains the arithmetic unit, internal storage unit, and control functions.

MAIN STORAGE — *See* Internal storage.

MATRIX PRINTING — A method of printing characters and other data by a pattern of small dots.

MEDIUM — The physical substance upon which data are recorded (e.g., diskette, magnetic tape, etc.).

MEGAHERTZ — Millions of cycles per second.

MEMORY — That part of a digital computer which stores data.

MENU — A display of selections that may be chosen, typically on a visual display screen.

MICROCOMPUTER — A digital computer made by combining microprocessor with one or more memory circuits. Single-chip microcomputers are also available.

MICROELECTRONICS — The field that deals with techniques for producing miniature circuits (e.g., integrated circuits, thin-film techniques, and solid-logic modules).

MICROINSTRUCTION — The most basic operation that takes place in a digital computer.

MICROPROCESSOR — The complete central processing unit for a digital computer (arithmetic-logic unit, control section, and some registers) on a single silicon chip.

MICROSECOND — One millionth of a second.

MILLISECOND — One thousandth of a second. Abbreviated ms.

MINICOMPUTER — A small and relatively inexpensive digital computer.

MISTAKE — A human failing (e.g., faulty arithmetic, incorrect keypunching, incorrect formula, or incorrect computer instructions).

MNEMONIC — A memory aid such as an abbreviation or acronym.

MODEM — A contraction of modulator/demodulator. Its function is to interface with data processing devices and convert data to a form compatible for sending and receiving on transmission facilities.

MONITOR — A television receiver without the circuitry to detect transmissions. Used widely with microcomputers.

MULTIPLEXER — A combinational circuit that applies the logic state of one of several inputs to a single output.

MULTIPROCESSING — Independent and simultaneous processing accomplished by a computer configuration consisting of more than one arithmetic and logic unit, each being capable of accessing a common memory.

MULTIPROGRAMMING — Pertaining to the concurrent execution of two or more programs by a computer. The programs operate in an interleaved manner within one computer system.

NANOSECOND — One billionth of a second. Abbreviated ns.

NEGATIVE LOGIC — A logic system where the binary bit 0 is represented by a high-voltage level and the bit 1 by a low-voltage level.

NETWORK — The interconnection of a number of points by data communications facilities.

NONVOLATILE STORAGE — A memory system that retains data without the need for electrical power.

NUMBER — The representation of a quantity. In digital computers, numbers can represent data, characters, instructions, and so on.

NUMERIC PAD — A special cluster of keys allowing input of the numeric digits 0 through 9.

OBJECT PROGRAM — A program written in or expressed in machine language.

OCTAL — A numbering system using base 8.

OFF-LINE — Peripheral units which operate independently of the central processing unit.

ON-LINE — Peripheral devices operating under the direct control of the central processing unit.

OPERATING SYSTEM — Software that controls the execution of computer programs and which may provide scheduling, input/output control, compilation, data management, debugging, storage assignment, accounting, and similar functions.

OPERATION CODE — The portion of an instruction that designates the operation to be performed by a computer (e.g., add, subtract, or move). Also called a command.

OUTPUT — Data transferred from the internal storage unit of a computer to a storage or peripheral device.

OUTPUT SECTION — A printer, video display, or other device that makes information processed by computer available to an operator or an electronic device.

OUTPUT UNIT — A device capable of recording data coming from the internal storage unit of a computer (e.g., card punch, line printer, CRT display, magnetic disk, or teletypewriter).

PARALLEL PROCESSING — Operating on data a chunk of bits at a time.

PARITY BIT — A binary bit added to a binary word to make the total number of 1s either even or odd.

PASCAL — A programming language that is of particular interest to computer scientists and is being used increasingly for many applications.

PATCH — To modify temporarily the software or hardware of a computer system.

PERIPHERAL DEVICES — Input/output and storage devices attached to a computer.

PERSONAL COMPUTER — A microcomputer with a keyboard input designed for ease of use and maximum economy.

PICOSECOND — One trillionth of a second. Abbreviated ps.

PL/1 — Programming Language1. A general-purpose programming language.

POSITIVE LOGIC — A logic system where the binary bit 1 is represented by a high-voltage level and the bit 0 by a low-voltage level.

PRECISION — The degree of exactness with which a quantity is stated. The result of a calculation may have more precision than it has accuracy. For example, the true value of π to six significant digits is 3.14159; the value 3.14162 is precise to six digits, given to six digits, but is accurate only to about five.

PRINTER — An output device that prints computer information on paper.

PROBLEM-ORIENTED LANGUAGE — A programming language designed for the convenient expression of a given class of problems (e.g., GPSS).

PROCEDURE — A precise step-by-step method for effecting a solution to a problem.

PROCEDURE-ORIENTED LANGUAGE — A programming language designed for the convenient expression of procedures used in the solution of a wide class of problems (e.g., FORTRAN, APL, PL/1, BASIC, and Pascal).

PROCESSOR — A digital computer.

PROGRAM — A list of instructions that tells a computer what to do and how to do it.

PROGRAMMING — The process of translating a problem from its physical environment to a language that a computer can understand and obey.

PROGRAMMING LANGUAGE — A language used to prepare computer programs.

PROM — Programmable read-only memory. A read-only memory, programmable by the purchaser, that cannot be erased.

PROMPT — A character(s) printed by the program to signal the user that input is required.

RAM — Read/write memory.

RANDOM-ACCESS MEMORY — A memory that offers equal access time to any storage location.

RAW DATA — Data that have not been processed.

READ — To sense data from a magnetic tape, disk, or punched card; or to make information in a memory available to some other circuit.

READ-ONLY MEMORY — A memory that contains permanent data which cannot be altered or erased. Often designated ROM.

READ/WRITE MEMORY — A memory which contains information that can be erased and modified. Often designated RAM.

READER — Any device capable of transcribing data from an input medium.

REAL-TIME SYSTEM — A system where transactions are processed as they occur.

RECORD — A group of related items of data treated as a unit (e.g., the inventory master record). A complete set of such records forms a file.

REGISTER — A string of flip-flops that stores one word of binary data. A register is a temporary memory.

RELIABILITY — A measure of the ability to function without failure.

RELOCATE — To move a routine from one portion of storage to another, and to adjust the necessary address references so that the routine, in its new location, can be executed.

REMOTE PROCESSING — The processing of computer programs through an input/output device that is remotely connected to a computer system.

REMOTE TERMINAL — An input/output device that is remotely located from a computer system. Also called a remote station.

RESOLUTION — The "fineness" of a visual display.

RESPONSE TIME — The time it takes the program or input/output device to respond to a user input or command.

ROM — Read-only memory. A type of memory permanently programmed by the manufacturer.

RUN — A single, continuous performance of a computer program.

SCAN — To retrieve or store data from beginning to end of a list or table.

SCREEN — The picture tube of a visual display terminal.

SCROLL — To display various sections of a long list of lines, similar to viewing a portion of a scroll as it is unwound.

SEMICONDUCTOR MEMORY — A computer memory that uses silicon integrated-circuit chips.

SEQUENTIAL LOGIC — A collection of logic gates that responds to incoming information only when a clock pulse is received. Sequential logic circuits use flip-flops so that each operation is affected by a previous operation.

SERIAL PROCESSING — Operating on data a bit at a time.

SOFTWARE — Paperwork such as programs and documentation associated with the operation of a computer.

SOLID STATE — The electronic components that convey or control electrons within solid materials (e.g., transistors, germanium diodes, and integrated circuits).

SORT — To arrange numeric or alphabetic data in a given sequence.

SOURCE COMPUTER — A computer used to translate a source program into an object program.

SOURCE DOCUMENT — An original document from which basic data are extracted (e.g., invoice, parts list, inventory tag, etc.).

SOURCE LANGUAGE — The original form in which a program is prepared prior to processing by the computer (e.g., FORTRAN or assembly language).

SOURCE PROGRAM — A computer program written in a nonbinary form such as assembly language or BASIC.

SPECIAL CHARACTER — A graphic character that is neither a letter nor a digit (e.g., the plus sign and the period).

SPECIAL-PURPOSE COMPUTER — A computer designed to solve a specific class or narrow range of problems.

STANDARD — An accepted and approved criterion used for writing computer programs, drawing flowcharts, building computers, and so on.

STATEMENT — The most elemental instruction to the computer in a higher-level programming language, such as BASIC or FORTRAN.

STORAGE — The retention of data so that the data can be obtained at a later time.

STORAGE DEVICE — A computer memory.

STORAGE LOCATION — A position in storage where a character, byte, or word may be stored.

STORAGE MAP — An aid used by the computer user for estimating the proportion of storage capacity to be allocated to data and instructions.

STORAGE PROTECTION — A device that prevents a computer program from destroying or writing in computer storage beyond certain boundary limits.

STORAGE UNIT — The portion of the central processing unit that is used to store instructions and data.

STRUCTURED PROGRAMMING — Techniques concerned with improving the programming process through better organization of programs and better programming notation to facilitate correct and clear description of data and control structures.

SUBROUTINE — A sequence of instructions in a computer program that is used more than once by the main program.

SYMBOLIC ADDRESS — An address expressed in symbols.

SYMBOLIC CODING — Coding in which the instructions are written in nonmachine language (e.g., a FORTRAN program).

SYNCHRONOUS — A computer operation that takes place under the control of a clock.

SYSTEM — An organized collection of machines, methods, and personnel required to accomplish a specific objective.

SYSTEMS ANALYSIS — The examination of an activity, procedure, method, technique, or business to determine what must be accomplished and how the necessary operations may best be accomplished by using electronic data-processing equipment.

SYSTEMS ANALYST — A person skilled in solving problems with a digital computer. He or she analyzes and develops information systems.

SYSTEMS PROGRAMS — Computer programs provided by a computer manufacturer. Examples are operating systems, assemblers, compilers, debugging aids, and input/output programs.

TELECOMMUNICATIONS — Pertaining to the transmission of data over long distances through telephone and telegraph facilities.

TERMINAL — A computer input/output device.

TEXT EDITING — The capability to modify text automatically.

THIRD GENERATION — Computers made with integrated circuits.

TIME SHARING — A method of operation whereby a computer system automatically distributes processing time among many users simultaneously.

TRACK — The path along which data are recorded, as in magnetic disks and magnetic drums.

TRANSLATE — To change data from one form of representation to another without significantly affecting the meaning

TRANSLATION — Conversion of a higher-level language, or assembly language, to machine-understandable form. *See also* Assembler, Compiler, and Interpreter.

TURNKEY SYSTEM — A computer system that includes all hardware and software to perform a specified application without the need of professional computer personnel.

TYPEWRITER — An input/output device that is capable of being connected to a digital computer and used for communications purposes.

VARIABLE — A quantity that can assume any of a given set of values.

VIDEO — Signals that generate display of data on a visual display terminal.

VIRTUAL MEMORY — A technique for managing a limited amount of high-speed memory and a (generally) much larger amount of lower-speed memory in such a way that the distinction is largely transparent to a computer user.

VOLATILE STORAGE — A memory system that retains data only when electrical power is present.

WORD — A string of binary bits used to represent a number, character, or instruction in a digital computer. Computer words can be any length.

WORD LENGTH — The number of bits, bytes, or characters in a word.

WORD PROCESSING — A text-editing system. A system that is used to prepare text, such as letters, forms, and pages of text.

WRITE — To place information into a memory or register.

B

Appendix

HEX-ASCII TABLE

	Control	Numeric	Uppercase	Lowercase	Special
00	Null	30 0	41 A	61 a	20 Space
01	Start of heading	31 1	42 B	62 b	21 !
02	Start of text	32 2	43 C	63 c	22 "
03	End of text	33 3	44 D	64 d	23 #
04	End of transmission	34 4	45 E	65 e	24 $
05	Enquiry	35 5	46 F	66 f	25 %
06	Acknowledge	36 6	47 G	67 g	26 &
07	Bell	37 7	48 H	68 h	27 '
08	Backspace	38 8	49 I	69 i	28 (
09	Horizontal tabulation	39 9	4A J	6A j	29)
0A	Line feed		4B K	6B k	2A *
0B	Vertical tabulation		4C L	6C l	2B +
0C	Form feed		4D M	6D m	2C ,
0D	Carriage return		4E N	6E n	2D -
0E	Shift out		4F O	6F o	2E .
0F	Shift in		50 P	70 p	2F /
10	Data link escape		51 Q	71 q	3A :
11	Device control 1		52 R	72 r	3B ;
12	Device control 2		53 S	73 s	3C <
13	Device control 3		54 T	74 t	3D =
14	Device control 4		55 U	75 u	3E >
15	Negative acknowledge		56 V	76 v	3F ?
16	Synchronous idle		57 W	77 w	40 @
17	End of transmission block		58 X	78 x	5B [
18	Cancel		59 Y	79 y	5C \
19	End of medium		5A Z	7A z	5D]
1A	Substitute				5E Λ or ↑
1B	Escape				5F _ or ←
1C	File separator				60
1D	Group separator				7B {
1E	Record separator				7C	
1F	Unit separator				7D }
7F	Rub out or delete				7E ~

Appendix

TIME UNITS

TIME UNITS AND THEIR FRACTIONAL EQUIVALENTS

Time Unit	Notation	Fractional Equivalent	Abbreviation
1 second	10^0	1 second	sec
1 millisecond	10^{-3}	1/1000 second	ms
1 microsecond	10^{-6}	1/1,000,000 second	μs
1 nanosecond	10^{-9}	1/1,000,000,000 second	ns
1 picosecond	10^{-12}	1/1,000,000,000,000 second	ps

Appendix
NUMBER-
CONVERSION
TECHNIQUES

Conversion of Numbers into the Base 10 System

To understand the theory of number conversion from one base to another, it will be useful to look at the meaning of a number in the decimal system.

Each decimal digit in a number represents a coefficient to be multiplied by the appropriate power of 10, the base. The position of the coefficient in the number string indicates which power of 10 is to be used in generating the products. The value of the number will be the sum of these products. Thus the number 125.64_{10} means

$$1. \times 10^2 + 2. \times 10^1 + 5. \times 10^0 + 6. \times 10^{-1} + 4. \times 10^{-2}$$

Any number in any base can be converted to its base 10 equivalent by summing the appropriate powers of the base, each multiplied by the proper coefficient. For example, 241.56_8 means

$$
\begin{aligned}
2. \times 8^2 &= 128. \\
+4. \times 8^1 &= 32. \\
+1. \times 8^0 &= 1. \\
+5. \times 8^{-1} &= .625 \\
+6. \times 8^{-2} &= \underline{.093\ 75} \\
&= \overline{161.718\ 75_{10}}
\end{aligned}
$$

$C4.A_{16}$ means

$$
\begin{aligned}
C(12.) \times 16^1 &= 192. \\
+4. \times 16^0 &= 4. \\
+A(10.) \times 16^{-1} &= \underline{.625} \\
& 196.625_{10}
\end{aligned}
$$

110.101_2 means

$$
\begin{aligned}
1 \times 2^2 &= 4. \\
+1 \times 2^1 &= 2. \\
+0 \times 2^0 &= 0. \\
+1 \times 2^{-1} &= .5 \\
+0 \times 2^{-2} &= 0. \\
+1 \times 2^{-3} &= \underline{.125} \\
& 6.625_{10}
\end{aligned}
$$

Conversion of Numbers from Base 10 into Other Bases

Conversion from base 10 into other new bases requires a two-part procedure. The first part of the procedure is used to convert the digits to the left of the radix point (i.e., the whole-number part). This is done by first dividing the base 10 number by the new base. This division may be accomplished using base 10 mathematics. The division process will yield an answer in the form of dividend and a remainder. The remainder becomes the first converted digit to the left of the radix point. Next, the dividend is divided by the new base to yield a second dividend and a new remainder which becomes the second converted digit. This process is repeated until the dividend is small enough to constitute only a remainder to become the final converted digit. The conversion of 74_{10} to base 8 would consist of

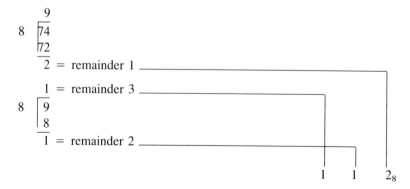

The second part of the conversion procedure is used to convert all the digits to the right of the decimal point (i.e., the fractional part). This conversion is done by multiplying the fractional part by the new base. The resulting product will be an integer plus a new fractional part. The integer is the first converted digit to the right of the decimal point, and the leftover fraction is used to generate the second converted digit by multiplying again by the new base. The result will be another integer plus a fractional part. This second integer is the second converted digit to the right of the decimal, and this second, leftover fraction is used to generate the third converted digit. The procedure continues until the leftover fraction becomes zero or until the desired number of significant digits is achieved. For example, $.55_{10}$ converted to base 8 will be

$$
\begin{array}{ll}
8 \times .55 = 4.40 & \text{first digit} = 4 \\
8 \times .40 = 3.20 & \text{second digit} = 3 \\
8 \times .20 = 1.60 & \text{third digit} = 1 \\
8 \times .60 = 4.80 & \text{fourth digit} = 4 \\
8 \times .80 = 6.40 & \text{fifth digit} = 6 \\
8 \times .40 = 3.20 & \text{sixth digit} = 3
\end{array}
$$

so that

$$.55_{10} = .43146\ 3146\ 3146 \ldots \ldots \ldots 8$$

repeating pattern

In this example we note that a repeating pattern appears. It is not always possible to find an exact equivalent for a fraction in two different base systems.

The main storage for a digital computer can store only a finite number of digits for a given fraction. It should also be mentioned that the digital computer does not round off answers in its printout. Rather, it is designed simply to truncate the values. Thus if the number $.55_{10}$ is given to the computer, converted to base 2, stored, retrieved from storage, and reconverted to base 10, the result may be truncated and printed as $.549999_{10}$.

Appendix

RS-232C INTERFACE CONNECTIONS

Pin Number	Definition	Signal Direction
1	Ground	–
2	Transmit data	to DCE
3	Receive data	from DCE
4	Request to send	to DCE
5	Clear to send	from DCE
6	Data set ready	from DCE
7	Ground	–
8	Data carier detect	from DCE
20	Data terminal ready	to DCE

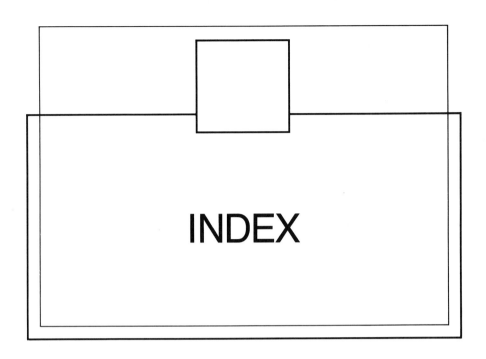

INDEX

SOFTWARE INFORMATION

BASIC software packages based on the algorithms in this text are available on 5¼-inch floppy diskettes for the APPLE II computer and for the IBM Personal Computer.

These packages contain a comprehensive user's manual and are available for purchase from:

Shoup Software
7382 San Sebastian Drive
Boca Raton, Florida 33433